MW01155542

Part of a Complete Breakfast

UNIVERSITY PRESS OF FLORIDA

Florida A&M University, Tallahassee
Florida Atlantic University, Boca Raton
Florida Gulf Coast University, Ft. Myers
Florida International University, Miami
Florida State University, Tallahassee
New College of Florida, Sarasota
University of Central Florida, Orlando
University of Florida, Gainesville
University of North Florida, Jacksonville
University of South Florida, Tampa
University of West Florida, Pensacola

Tim Hollis

University Press of Florida
Gainesville
Tallahassee
Tampa
Boca Raton
Pensacola
Orlando
Miami
Jacksonville
Ft. Myers
Sarasota

Part of a Complete Breakfast

Cereal Characters of the Baby Boom Era

Printed in the United States of America. This book is printed on Glatfelter Natures Book, a paper certified under the standards of the Forestry Stewardship Council (FSC). It is a recycled stock that contains 30 percent post-consumer waste and is acid-free.

17 16 15 14 13 12 6 5 4 3 2 1

LIBRARY OF CONGRESS CATALOGING-IN-PUBLICATION DATA
Hollis, Tim.
Part of a complete breakfast : cereal characters of the baby boom era / Tim Hollis.
p. cm.
Includes bibliographical references and index.
ISBN 978-0-8130-4149-0 (alk. paper)
1. Cereal products industry—United States. 2. Advertising—Breakfast cereals. 3. Television advertising and children—United States. 4. Popular culture—United States. I. Title.
TP435.C4H65 2012
664'.7—dc23 2012009833

University Press of Florida
15 Northwest 15th Street
Gainesville, FL 32611-2079
http://www.upf.com

Contents

Author Tim Hollis's museum in Birmingham, Alabama, is a valuable repository of popular culture memorabilia.

Preface

Opening the Box

Since the first question most authors face about any of their work is, "What made you think of doing a book about *that*?" it seems only fair to give credit where it is due. This book was not my idea.

A few years ago, I was in Gainesville for a marketing meeting at the University Press of Florida. I have done two other books for that fine company, and while I was in town, we were discussing some of the sales possibilities. Then-editor John Byram and I went to lunch. Over bacon cheeseburgers and steak fries, he told me he had had an idea for a book he thought needed to be done, and based on my past work—not only for UPF but also for several other publishers—he thought I might be the right author for it. Then came the big question: "Do you think you could come up with anything about all of those great old cereal advertising characters . . . Tony the Tiger, the Trix Rabbit, and all?"

My iced tea almost came out of my nose I was trying so hard to stifle a laugh. "Yeah, I think I might be able to say a thing or two about that," I said. What John did not know at the time was that in my own personal museum near Birmingham, Alabama, I had several displays of cereal character memorabilia mixed in with a few dozen other categories of baby boomer pop-culture history. I knew I would not have to go far to find the material I needed for his proposed volume.

It might seem strange that no one has ever tackled the topic before, or at least not in the way John and I had in mind. In the early 1990s, a gentleman named Scott Bruce was, for a while, the nation's leading expert in cereal memorabilia. He produced a newsletter appropriately named *Flake*, two pictorial volumes of cereal boxes and their prizes, and a heavy tome he called *Cerealizing America*. Those books laid the groundwork for this one but did not handle the material in quite the same way. The pictorial books were just that—no text—and his other book was meticulously researched but was, in the end, a history of the cereal industry that touched on the advertising characters only as part of the huge overall story.

So, here we are with this new approach to the topic. John Byram has since moved on to other pastures, but I stuck it out to get this project done, even if it sometimes meant scraping up the little crumbs that settle at the bottom of the cereal box. Along with tracing how the various ad icons evolved over the years, wherever applicable I have tried to point out how they reflected the particular period in which they were created. This is often a quite subjective topic, and to be truthful, the most successful of the cereal characters were the ones that were *not* tied to any current trend or fad. That is what has enabled Tony the Tiger to survive in the advertising jungle for sixty years without changing his stripes, for example.

One word about the illustrations in this book: most of them, with exceptions noted in the credits at the end, came from my own collection. In the years since Scott Bruce's seminal works, dozens of vintage cereal commercials have been discovered and preserved on videotape, DVD, and YouTube. Unfortunately, the quality of these reproductions ranges from mint condition to horrible. Frame grabs from the actual commercials have been used when there was no other way to illustrate a particularly important moment in advertising character history, but be forewarned: some of them look better than others. A few look almost like 8 × 10 glossies, such is the clarity, while others look much as all of us first saw them as kids: beamed by a TV station many miles away and picked up on a small tabletop set with a rabbit-ears antenna. In the case of those less-than-perfect images, just show them to your kids and make them understand how lucky they are to be living in the days of big-screen TVs and high definition.

Then again, maybe we were the lucky ones. We did not have all that technology, but we had the magical realm known as Saturday-morning cartoons—and many times the commercials were even more entertaining than the programs they interrupted. As you make your way through the following pages, do not be surprised if you suddenly have an inexplicable craving for Frosted Flakes, Lucky Charms, or Cap'n Crunch. That is what these characters were designed to do—and many of them can still put us under their commercial spell to this day. Got the milk? Got a bowl? It's time to start pouring on the history.

Breakfast Pals

The World before Television

Attempting to come up with some sort of chronology for the towering family tree of cereal characters can be somewhat like digging a hole at the sandy beach. Every shovelful of matter that is cleared away is immediately replaced by another batch rushing in to take its place. As we will see repeatedly as we move along, for most of this story, many developments were taking place simultaneously—some intentionally and some not. If we are to begin at the beginning, however, we have to travel back to the birth of the cereal industry in late-nineteenth-century America.

Oops, here comes another rush of sand. When speaking of the cereal industry as it relates to the advertising characters we all grew up watching on television and gazing at on boxes as we wolfed down our breakfasts, the natural inclination is to think of cold cereal, usually half-drowned in milk. Cold cereal, however, was a later innovation; the first cereals to be marketed were of the hot variety—which is where the first-pitch characters came into the picture.

Historians have traced the birth of the cereal industry to 1854, when immigrant Ferdinand Schumacher started a small company to grind oats into oatmeal. While this dish had been a breakfast staple in Scotland, the vast majority of Americans considered oats to be strictly horse fuel.

The Quaker Oats man, dating to the late 1870s, became the United States' first trademarked character relating to breakfast food.

Schumacher had to put up with newspaper writers merrily accusing oatmeal eaters of developing a whinny, but he got the last horse laugh on them when an influx of immigrants in the 1870s made his product a success beyond anyone's imagination.

By all accounts, Schumacher was an austere, no-nonsense businessman who scowled at the very idea of advertising—another industry that was just beginning to get past the crawling stage in the years after the Civil War. In 1877, one of his competitors, Henry Crowell of Ravenna, Ohio, began to fight back by developing a brand name for his own line of oatmeal. He called his product Quaker Oats and decorated the cardboard containers with a rather lean William Penn look-alike who grasped a scroll bearing the single word "Pure." When the Quaker Mill Company registered the "figure of a man in Quaker garb," as the official description read, he became the first trademarked character in cereal history.

Because characters and figures such as the Quaker Oats man and his many descendants seem so much like old friends, having hung out in our supermarkets and kitchens for as long as any of us can remember, it is sometimes easy to forget that at their root they are simply trademarks. In his excellent book *Symbols of America* (1986), historian Hal Morgan takes the reader on a fascinating journey through the development of brand name and trademark law. Morgan traces the concept as far back as medieval times: "Potters' marks were joined by printers' marks, watermarks on paper, bread marks and the marks of various craft guilds. In some cases, these were used to attract buyer loyalty to particular makers, but the marks were also used to police infringers of the guild monopolies and to single out the makers of inferior goods." Morgan reports that this early concept of trademarks and brands made the journey across the Atlantic when the first Europeans brought their Old World traditions to the New World. However, since we know the pilgrims were not serving corn flakes endorsed by Tony the Turkey at their first Thanksgiving dinner, we must turn again to Morgan to see just when the big turnaround came:

> It was not until the nineteenth century, when dramatic changes occurred in the way goods were made and sold, that trademarks and package labels grew into a strong, positive factor in the American economy. As individually packaged goods replaced bulk-packaged merchandise, manufacturers found a new sign board—the wrappers, boxes and labels—on which to promote their names to customers. And as the practice of advertising matured, industry found itself selling brand names directly to the public, rather than hoping the retail merchant would do the job.

It might be amusing to note that one of the oldest food trademarks still in use, registered with the U.S. Patent Office in 1870, was the opposite in many ways from the Quaker man that would soon follow: the horned devil of William Underwood & Co. of Boston, "for use on Deviled Entrements." His satanic visage, though modernized, continues to leer from packages of Underwood's deviled ham today.

In the ensuing decades after the beatific Quaker man joined the Underwood devil on store shelves, Schumacher's and Crowell's oatmeal empires would merge with others until the giant Quaker Oats Company emerged triumphant. The Quaker man was front and center in all of the firm's advertising, even at one point appearing in a mural-sized ad painted on England's beloved White Cliffs of Dover. The Quaker man did not have a personality except for the broad characteristics associated with the Quaker church; as a matter of fact, the oatmeal company and the church maintained an uneasy relationship from the very beginning, as commercialism found itself unequally yoked with spiritual health. (A particularly interesting clash of the two directly involving an advertising character would take place in the late 1980s, but that is a story for a much later chapter.)

From his original solemn design, the Quaker man was redesigned in the mid-1940s by commercial artist Jim Rich, and a decade later was further refined by Haddon Sundblom, whose primary claim to fame is his rendition of the classic Coca-Cola Santa Claus. Sundblom's Quaker beamed broadly as a full-color painting on the front of every package of oats, and his smile can still be seen there today. The figure occasionally ventured into the broadcast media, proclaiming in a deep voice, "Nothing is better for thee than me." Yes, the Quaker Oats man was, in every way, the grandfather of all the cereal characters who would follow.

If oats were being milled into a hot cereal for breakfast consumption, could wheat be far behind? In 1893, miller Tom Amidon began marketing a breakfast food he named Cream of Wheat. Strictly as a way to enliven the crude, early packaging, his business partner Emery Mapes found an old printer's block depicting an African American chef with a saucepan over his shoulder. Cream of Wheat was an immediate success, and its logo chef was soon improved upon by various commercial artists, who produced ad after ad showing the jolly figure in a variety of situations.

The Cream of Wheat chef might, at least in some circles, today be considered an ethnic stereotype—but if so, he certainly was not a negative or derogatory one. True, like the Quaker Oats man, he was not required

This 1918 advertisement depicts the Cream of Wheat chef in his most familiar appearance. Unlike his fellow African American ad characters, Aunt Jemima and Uncle Ben, the chef was such a universally appealing figure that he has not required updating to keep up with the times.

to display any personality other than being genial. It is significant to note, though, that the chef continues his long-established appearances on Cream of Wheat boxes in the twenty-first century, and has not had to undergo the drastic makeovers his fellow logo characters Aunt Jemima and Uncle Ben have experienced to align them with modern sensibilities.

While the Quaker man and the chef were pushing their hot and hearty breakfast wares at the public, a craze for cold (or, to use the more official name, ready-to-eat) cereals was being birthed in Battle Creek, Michigan. The story of how this city became the incubator for two of the future's largest cereal companies has been told in a number of books and was even the subject of the 1994 film *The Road to Wellville*. To summarize the more than twice-told tale, the Battle Creek Sanitarium was operated by the Seventh-Day Adventist church, so the residents who came from far and near were required to adhere closely to the church's vegetarian diet. The head doctor at the sanitarium was John Harvey Kellogg, who spent much of his free time in the facility's kitchens, trying out formulas for new foods that would be healthful and tasty while remaining within the church's dietary guidelines. John Harvey's younger brother, W. K. Kellogg, was more interested in the commercial possibilities inherent in their joint work in the cereal industry. So was one of the patients at the sanitarium, C. W. Post.

In 1894, the Kellogg brothers invented a cereal made from toasted corn flakes, and while they were having a fraternal squabble as to whether they wanted to commercialize the product (guess which brother was in favor of what), ex-patient Post beat them to the store aisles with a similar product he called Elijah's Manna. Inasmuch as the boxes featured artwork of the famed prophet, it is likely Elijah was the only Old Testament figure to date to serve as a cereal mascot. Even his august presence was not enough to move the product, and the cereal soon received its new name, Post Toasties.

Finally, in 1903, W. K. Kellogg prevailed over his more conservative brother and began marketing their invention as Kellogg's Corn Flakes. The company has apparently always had confidence in Corn Flakes to sell itself rather than relying on cartoon ad characters, except for a brief time in the late 1950s and 1960s, about which we will hear more later. Around 1907, the Corn Flakes packaging began featuring artwork depicting a beautiful young girl known as the Sweetheart of the Corn, but like her male counterparts in the hot cereal business, she was not a character in the usual sense of the word. Regardless, she would also be revised, mod-

The Sweetheart of the Corn was the first advertising character for a Kellogg's cereal. She first appeared on boxes of Corn Flakes around 1907.

ernized, and updated over the next fifty years, no matter how corny some ad executives might have considered her to be.

While Kellogg's was just beginning to sow its wild oats—or corn, or whatever grain they were experimenting with—one of the company's new competitors was poised to introduce a figure that now stands as the most likely candidate for the first true cereal advertising character with personality, as opposed to being simply a logo. In 1902, a new cereal named Force—basically a wheat clone of Corn Flakes—made its debut with the help of a cartoon character known as Sunny Jim.

Sunny Jim promoted Force through a series of print ads—the only type of advertising possible in those prebroadcasting days—created by two young women, writer Minnie Maud Hanff and artist Dorothy

Sunny Jim, the mascot for Force cereal, received his zest for life by generous consumption of his product. This particular ad was part of a series created by famed *Wizard of Oz* illustrator W. W. Denslow.

Flicken. The short, short stories consistently featured the morose figure of Jim Dumps, who hated everything in life. Eating a bowl of Force cereal had a magical effect on Dumps, who would be transformed into the ever-optimistic Sunny Jim by the tale's end. These stories were eventually collected into giveaway premium booklets, and after Flicken's departure from the campaign, the artwork duties were assumed by W. W. Denslow, primarily famed as the original illustrator of *The Wizard of Oz* (1900).

The Sunny Jim campaign became legendary in the advertising world but mostly for the wrong reasons. Professional ad executives hated the campaign upon its debut and spent much of the ensuing fifty years trying to tell anyone who would listen that it illustrated the absolutely wrong way to go about selling a product. Further research has proven that Sunny Jim was, in fact, an extraordinarily successful salesman when it came to moving Force cereal. Historian Eileen Margerum itemized the reasons for the disconnect between perception and reality in this particular case:

> The advertising craft was afraid of what the success of the Sunny Jim campaign represented. Advertising was still struggling to be taken seriously as a reputable and necessary part of American business culture. This campaign challenged all the claims to special expertise being made . . . It was written and drawn by two "girls" . . . In addition, members of the public wrote their own Sunny Jim jingles as if they, too, could write ad copy.

For all the argument about it, the Sunny Jim stories found even greater success in England than in the former colonies. In fact, the cereal and Sunny Jim have both survived to be a major part of the British cereal industry to this day. On this side of the big pond, Sunny Jim was all but forgotten by 1933, when the cereal company became the sponsor of a children's radio show concerning a young cowboy named Bobby Benson and his guardian, good old Sunny Jim. After less than a year, Jim had been dropped from the cast, but the show thrived for the rest of radio's golden age.

Sunny Jim's brief excursion into the medium notwithstanding, in the early 1930s radio began to give serious competition to all previous forms of advertising. The sheer power of broadcasting to millions of listeners at the same time soon began influencing the way advertisers looked at their potential consumers, and as with the case of the *Bobby Benson* show, the children's audience suddenly became a highly desirable demographic.

Kellogg's had continued to plow fresh ground ever since the initial success of Corn Flakes, and by the late 1920s the company had produced a number of additional cereal products, most of which are long since forgotten. One of them, introduced in 1928, did survive to become the company's second-most famous product: Rice Krispies. At first, no one seemed to have given any more thought to tying this product to a cartoon character than Kellogg's previous cereals, but radio was to irrevocably alter that course.

In 1931, Kellogg's began sponsoring a children's radio show known as *The Singing Story Lady*, starring Ireene Wicker. As the eponymous host, Wicker recited nursery rhymes and sang simple children's songs between dispensing advice to eat Kellogg's cereals. A giveaway booklet from the program's first year describes Rice Krispies as "crunchy rice bubbles that actually crackle out loud." Shortly thereafter, the ad executives came up with an additional pair of onomatopoetic words, and Ireene Wicker began plugging Rice Krispies as the cereal that went "snap, crackle and pop."

Illustrator and commercial artist Vernon Grant had already won some renown for his depictions of elves, gnomes, brownies, and other fairy-tale fauna, and someone (accounts differ) decided to have him illustrate a trio of elfin figures to personify the three sounds made by Rice Krispies. Incorporating exclamation points into each name, the public was soon introduced to "Mr. Snap!," "Mr. Crackle!," and "Mr. Pop!" Eventually the honorific titles were dropped, and Snap!, Crackle!, and Pop! became Kellogg's longest-running ad characters.

This is not to say the noisy trio did not undergo some major changes along the way. As Vernon Grant originally drew them, there was little to distinguish Snap! from Crackle! from Pop! The three were identical in their physical features and of course had no need for any separate personalities or depth of characterization. Even the outfits they wore were not standardized, although one might say that Snap! generally resembled a baker, Pop! a soldier, and Crackle! a sort of generic elf. By the end of the decade, they had snapped up another place in history by becoming the first cereal characters to be adapted into animated form, in a ninety-second spot meant to be shown among the other selected short subjects in movie theaters.

Breakfast Pals, as the film was titled, was produced in either 1937 or 1938—like the question of who thought up the idea for the characters, it depends upon what source one consults. It opens on a morning shot of two semi-realistically animated boys at the breakfast table; considering that even such major players as the Walt Disney Studio were still struggling to convincingly animate believable humans, it is somewhat remarkable that this film succeeds as well as it does. "Gee, I'm glad ya stayed all night, Bobby," says one of them as the other picks listlessly at his bowl of gray, particularly unappetizing cereal. "Don'tcha like the cereal?" pointlessly asks the host kid.

"Well, yeah," lies Bobby, "but it's a little mushy . . . it's not crisp like the kind my breakfast pals serve my dad and me at home." With a whistle, he summons Snap!, Crackle!, and Pop!, who look absolutely nothing like their appearances in Vernon Grant's ads, all three wearing chef's hats. No sooner have the good guys shown up than their evil counterparts (Soggy, Mushy, and Squishy) materialize out of the competition's cereal

This ad from 1933 shows the original Vernon Grant designs for Kellogg's trio of Rice Krispies mascots: Snap!, Crackle!, and Pop!

In 1937–38, Snap!, Crackle!, and Pop! became the first cereal mascots to be featured in animation, in a ninety-second adventure shown along with other short subjects in movie theaters. Their design looked nothing like the original Vernon Grant version.

box. "Hey, gang, dese guys is musclin' in," snarls one of the toughs. "Let's get 'em."

A typical, if brief, cartoon battle ensues, which ends with Snap!, Crackle!, and Pop! rolling up their gangster rivals in pancakes and dousing them with syrup. Only in the closing seconds of the film do they step aside to reveal the Rice Krispies box in all its glory; presumably the audience was only then finding out that what they had been watching all along was a commercial.

(Advertisements in film-industry trade publications of the era indicate that three other Rice Krispies cartoons were produced: *Breakfast Harmony*; *Sinking, Sinking, Sunk*; and *Pantry Purge*, which sounds like a cautionary tale about anorexia . . . but no doubt was not. Prints of these productions are not yet known to have surfaced.)

Meanwhile, back in the radio world that had inspired the three audio-based elves, more and more cereal companies were finding creative ways to tie their products in to the successful children's shows of the era, in effect turning those programs' leading characters into advertising spokespeople. Two such series debuted in 1933: *Jack Armstrong, the All-American Boy* and *The Tom Mix Ralston Straightshooters*.

As the title would indicate, the latter of the two was backed by Ralston-Purina (the company made it a policy to use the Ralston brand for human food and the Purina brand for anything not intended for consumption by people). *Jack Armstrong* was a product of General Mills, a newcomer to the cereal scene that had been formed by a merger of several smaller companies in 1928. General Mills wanted to push its wheat flakes known as Wheaties, immodestly touted as the "Breakfast of Champions." What better champion could there be than a high school athlete who excelled in every competition? That was the starting point for young Master Armstrong, though within a short time his adventures would carry him so far around the globe that concerned parents actively began questioning what had become of his scholastic career. Regardless, General Mills' commercial jingle managed to tie the name of the product and its main spokescharacter into a single memorable song:

> Have you tried Wheaties? They're whole wheat with all of the bran;
> Won't you try Wheaties, for wheat is the best food of man.
> They're crispy, they're crunchy the whole year through,
> Jack Armstrong never tires of them and neither will you!
> So just buy Wheaties, the best breakfast food in the land.

While Jack was champing down on the breakfast of champions, Ralston was going west, young man, with the fictional adventures of a real-life character: rodeo and movie cowboy Tom Mix. By all accounts, Mix had virtually nothing to do with the radio show bearing his name—even his titular role was played by other actors—and thus, when he died in 1940, the show was able to keep going as if Mix had been as fictitious as Jack Armstrong. One of the jobs the faux Mixes had to perform was the theme song, which, like General Mills' masterpiece, demonstrated the product's tie-in to the star:

> Shredded Ralston for your breakfast
> Starts the day off shinin' bright.
> Gives you lots of cowboy energy
> With a flavor that's just right.
> It's delicious and nutritious,
> Bite size and ready to eat.
> Take a tip from Tom, go and tell your mom
> Shredded Ralston can't be beat!

Both Wheaties and Ralston (which came in both a hot and a cold version) occasionally used their boxes to advertise their radio offerings, if not quite to the extent that television would make series characters into advertising mascots. Besides the jingles, the primary connection between product and star was in their ingenious use of premiums that would be introduced into the continuing storyline and then offered to juvenile listeners for a box top or two. No one who has been paying attention for the last forty years needs to be told these seemingly throwaway doodads now command extravagant prices on the collectors' market, simply because so relatively few of them were considered important enough to preserve.

Another hero sailed into radio in 1935, and his program would have even more far-reaching effects on cereal advertising than the likes of Jack Armstrong and Tom Mix. Popeye had been a newspaper comic-strip staple since 1929, and his long-running series of animated cartoons was launched in 1933. It was no surprise for him to also transition into radio, but the *Popeye the Sailor* series had some characteristics that set it apart from either of the one-eyed mariner's other media.

The voices in the radio show were reminiscent of the cartoon voices but not very close duplications. There was apparently no Bluto to serve as a continuing villain, either. Instead of infant Swee'pea, for purposes of radio Popeye had an "adoptik kid" of elementary school age, whom he called Matey. (Obviously it would have been rather useless in an audio-only medium to have a major character who could not speak.) But the biggest departure from any previous version of Popeye was due to the sponsor. Wheatena was a hot cereal that had more in common with Cream of Wheat than with Wheaties, and in the radio series—you might want to sit down for this one—there was no mention of Popeye's spinach, as he obtained instantaneous super strength from ingesting various numbers of bowls of Wheatena.

Only a handful of recorded episodes of the *Popeye the Sailor* radio show have survived, and it would do our discussion well to examine at least one of them in some detail. Each show opens with the sound of a ship's bell, running feet, and the announcer's call, "All hands on deck! Here's Popeye!" The orchestra then strikes up the famous "I'm Popeye the Sailor Man" theme song, but the announcer takes up the cadence with a new verse: "Wheatena's his diet, he asks you to try it/With Popeye the Sailor Man!"

The particular episode under our microscope here opens with the announcer commenting on Popeye's fan mail. According to our emcee, it seems all the boys want to be football players, so they eat Wheatena

Hot cereal Wheatena sponsored the *Popeye the Sailor* radio series beginning in 1935. Instead of eating spinach, Popeye gained instant "muskle" by ingesting bowls of the cereal he just happened to be carrying around with him.

because it "makes muscles." All the girls "want to grow up in a hurry and be young ladies," so they eat Wheatena because of "the roses it puts into their cheeks." Thus is set up an obvious double standard that would still be plaguing cereal advertising well into the 1980s.

As the plot gets under way, Popeye, Olive, Matey, and Wimpy are visiting the zoo and commenting on the odd animals. Seeing a camel, Matey observes, "Gee, Popeye, those bumps on his back look just like your muscles!" "Arf arf arf!" laughs Popeye. "If me muskles was that big, I could never get me shirt on." Then they notice this zoo features elephant rides as an attraction, and Matey wants to take advantage of the opportunity. With Matey aboard the elephant's back, Popeye insists that the reluctant Olive go along for the ride. Everything goes swell until the elephant is frightened by a mouse that wanders into the cage ("Greetings, Mickey," mumbles Wimpy). The pachyderm is thrown into a frenzy and begins to stampede, with Olive and Matey hanging on for dear life.

Popeye tries to catch the running elephant but only gets trampled for his efforts. He tells Wimpy that in order to rescue Olive and Matey, he is going to need four bowls of Wheatena. (They do not explain how Wimpy could have four bowls of already-prepared hot cereal on his person any more than the cartoons explored how Popeye could conceal a metal can of spinach inside his shirt.) Popeye gobbles down the cereal and growls, "NOW look at me muskles!" Here is where radio's role as "theater of the mind" comes into play, as the audience is left with a musical sting by the orchestra and a collective whistle by the other patrons of the zoo to picture Popeye's flex. About that time, the elephant grabs Popeye with his trunk and tries to smash him against the bars of the cage, but Popeye delivers a humdinger of a punch and knocks the beast out cold.

Everyone celebrates Popeye's victory—everyone except Popeye. "I yam a sad swab," he says, "on accounta I had ta sock a poor dumb aminal." "Oh, don't worry about the elephant," says the zookeeper. "He's got a thick skull; he'll get over it." "I yam glad," replies Popeye, "and since I did not means ta hurt him, I hopes this elephink forgets." He then comments on the pandemonium caused by a tiny mouse and delivers the story's moral: "Never underestimate the *little* fellow."

Although the *Popeye* radio show ran for only a couple of years, this substitution of Wheatena for spinach would have far-reaching effects in cereal advertising. Even the makers of Wheatena seem to have retained a vivid memory of it and eventually decided to rectify that 1930s-era double standard mentioned above. In a circa 1970 TV commercial, a man is shown amorously fondling a beautiful, aristocratic young lady and admonishing her, "I think it only fair to warn you, I had a bowl of Wheatena this morning." Without a word, the lady grabs his arm and throws him to the ground with a quick judo move, then stares coolly at the camera as she replies, "So did I." Other cereal companies would take a few years longer to catch on to those changing attitudes.

It would take a team of psychologists to explain just why the concept of a food product endowing instantaneous superpowers has been such a staple of pop culture ever since Popeye popped open his first spinach can or swallowed several bowls of Wheatena consecutively. Back in 1975, cartoon memorabilia collector Robert Lesser dissected what he believed to be the source of this success:

> Part of Popeye's general appeal was linked to the social ethos of the 1930s: Kids were manlike if they could use their fists, and dads encouraged their young males to stand up and fight the other kids on the block. Prizefighting and wrestling were popular and accepted sports then. Popeye was always ready to fight and at the beginning would take severe punishment before the can of spinach was gulped. In the language of the streets, "he could take it like a man."

A few years later, Bud Sagendorf—who served as assistant to Popeye's creator, E. C. Segar, and later took over drawing the newspaper strip himself—further speculated that Popeye was a natural product of the Great Depression: "A frustrated population liked the idea of one small man fighting back and winning. They, too, wanted to strike out at something they feared and didn't understand." Such analyses probably have more than a cereal flake of truth but do not explain why the concept remained popular long after the Depression was just something parents and grandparents talked about. It is true, however, that when such once-glorified violence became less socially acceptable, the advertising staple of "eat this and you will be able to destroy your opponent" went away—for a little while, at least.

There is no doubt that radio was a major outlet for cereal advertising, but because it lacked visuals, it was not much of a spawning ground for original mascot characters. That did not mean the sponsors of popular shows could not tie themselves to the regular cast of characters, but it was usually in a slightly different way than the Popeye series' method of incorporating the product into the script. One of Kellogg's less-remembered cereals, Pep, was the long-running sponsor for the *Superman* radio show. Although the announcer, Jackson Beck, constantly referred to the product as "the super-delicious cereal," there was no implication that eating it gave Superman his powers—although there was some hinting that eating it might produce similar strength in its consumers, as long as they were boys instead of girls. Pep did, however, get into the premium business in a big way, and Kellogg's would maintain its connection with Superman well into the television era.

Lest one might think that only children's radio programs pushed cereal, let us briefly consider the case of Jack Benny. The thrifty comedian had long been sponsored by Jell-O, but when sales reached a critical mass,

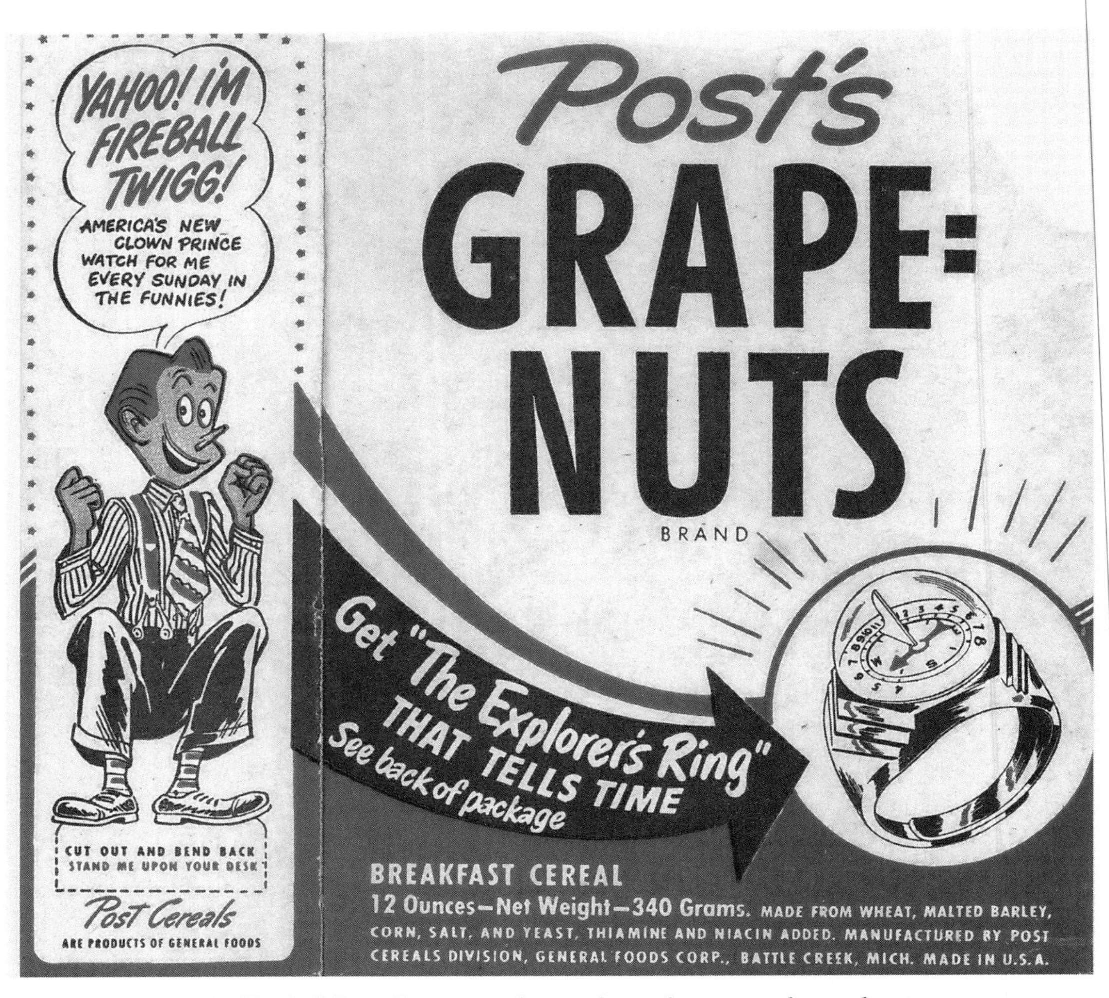

Post's "Fireball Twigg" was one of several cereal mascots whose adventures were confined almost exclusively to the Sunday comics sections of local newspapers. Occasionally, as with this example, they would be permitted into the side panels of the cereal boxes, but they rarely made it onto the front.

General Foods decided to switch Benny to one of its other products, Post's Grape-Nuts and Grape-Nuts Flakes. The comedian's portrait is known to have appeared on boxes of both cereals during the early 1940s, but his primary role was to continue his established practice of humorous commercials gently spoofing his own sponsor. Unlike the children's programs, shows aimed at the prime-time adult audience did not indulge themselves with premium offers nearly as often. There is no record of any Jack Benny premiums offered by the Post products, although there had been a handful of them during his Jell-O years.

During the 1940s, the major medium for introducing original advertising mascots was the advertising appearing in such weekly magazines as *Life* and the *Saturday Evening Post*, coupled with special artwork shoehorned into the Sunday comics section of local newspapers. Ad executives might have taken a perverse joy in running down the Sunny Jim campaign of yore, but they were hardly bashful when it came to utilizing the same storytelling technique on their own.

Cream of Wheat must have realized there was not much they could do with the adventures of their trademarked chef, so around 1940 the company struck a deal with Al Capp, creator of the immensely popular *Li'l Abner* comic strip. For well over a decade, magazines carried specially drawn *Li'l Abner* strips touting the benefits of eating Cream of Wheat. The miniature adventures seemed equally influenced by Sunny Jim's escapades and Popeye's more recent experiences with Wheatena. Each story presented some sort of quagmire or another, which the brawny Li'l Abner was ill-equipped to handle until Mammy Yokum cooked up a five-minute bowl of Cream of Wheat. After consuming the meal, Abner would have more than enough of "that Cream o' Wheat feelin'" to single-handedly beat off man or beast, or otherwise settle his adversary's hash. The Cream of Wheat chef would appear only at the bottom of each ad, sometimes using his limited space to drive home some hard-selling facts about the product.

While all of this was going on in Dogpatch USA, the real, live chefs in General Mills' kitchens were not sitting around on their Wheaties. In 1941, they introduced their other longest-running cereal success, a boxful of tiny circular O's they dubbed Cheerioats. Yes, that was its name at the time, and it made the fine folks at Quaker Oats quake in their buckled boots. Quaker took General Mills to trademark court, claiming they owned the exclusive right to use the word "oats" in their product names. This might have been open to some debate, but General Mills was not interested in fighting over their oat O's. Instead, they decided "Cheerios" was a better name, and more descriptive of the shape of the cereal, so Cheerios they have remained ever since.

The logo character for the original Cheerioats, appearing in comic strip–style ads and occasionally on the box panels, was a miniature girl called Cheeri O'Leary. Had she been adapted into animated form, she likely would have been given an Irish accent to go along with her name, but in the ads she did little beyond helpfully pointing out the joys of her tiny oat rings. In one episode from 1943, Cheeri cheers up western movie

comedian Andy Devine, who cannot muster the vocal strength to call his hogs. A bowl of Cheerioats restores his familiar gravelly tones enough to bring the swine in from near and far.

(The inclusion of Devine in this particular ad is historically interesting. For one thing, at the time it was published, he was appearing semi-regularly on Jack Benny's radio show, sponsored by rival Grape-Nuts. Also, Devine would be playing a huge role in Kellogg's cereal advertising on television for a decade beginning in the mid-1950s. General Mills' snagging him for a Cheerioats endorsement only seems to prove he was in high demand by cereal manufacturers of all types.)

Another elfin figure was Betty Bite Size, who handled Shredded Ralston selling chores when the fake Tom Mixes were not on the job. With doll-like eyes, pigtails, and a dress with Ralston's famous checkerboard pattern,

In 1949, Sunday comics featured the escapades of Cheerios Joe, a young do-gooder who was the more-or-less direct ancestor of the product's long-running Cheerios Kid.

Betty Bite Size was just too cute for words. What some readers of today would NOT find so cute about her ads was their mercifully outdated attitudes toward husbands and wives.

As an example, let's take one Sunday comic supplement ad with the title "This Bride Woke Up Just in Time." It begins with a husband trying to get his lingerie-clad wife out of bed: "Aren't you gonna get up for breakfast, honey?" "Goodness, no! Breakfast is such a dull meal," replies his laconic mate. The next panel reveals this cad's innermost thoughts as he mopes, "Humph! Thought when I got married, I'd have a wife who'd fix me a good breakfast." Betty Bite Size happens to be watching, and instead of hitting the lazy lout upside the head and telling him to make his own breakfast, she rushes to get the bride out from under the covers before it is too late.

A DELICIOUS NEW OATMEAL CEREAL

with CORN and RYE added!

Here it is, folks! America's *old reliable* for breakfast nourishment, in a delectable ready-to-eat form! Cheerioats comes in crisp little "doughnuts," filled with flavor and nutritive value! For it's 75% ground oatmeal, with corn and rye added for crispness and flavor! Actually provides *full oatmeal amounts* of Vitamin B_1, Vitamin G, Calcium, Phosphorus, Niacin and Iron! Try a bowlful of CHEERIOATS for breakfast tomorrow! GENERAL MILLS, Inc., Minneapolis, Minnesota.

"YEP FOLKS, FINE IS THE WORD FOR CHEERIOATS! FINE FOR FLAVOR AND FOR OATMEAL GOODNESS!" SAYS ANDY DEVINE

Now Starring in Howard Hawks' "CORVETTES IN ACTION" A Universal Picture

NEEDS NO COOKING!

Copyr. 1943, Gen. Mills, Inc. CHEERIOATS is a reg. trade mark of Gen. Mills, Inc.

The original name for General Mills' tiny oat rings was Cheerioats. In this 1943 comic-strip ad, mascot Cheeri O'Leary uses the power of a nutritious breakfast to help movie and radio comedian Andy Devine.

The indescribably cute Betty Bite Size was used in advertising for two of Ralston's products, the hot Shredded Ralston and the cold cereal Chex. This ad introduced the new Rice Chex in January 1951.

By 1948, Snap!, Crackle!, and Pop! had received a major makeover and looked much as they do today. If you look closely at this ad, you can see the original Vernon Grant versions of their heads were still appearing on the Rice Krispies boxes, though.

Maybe it's already too late, as the grumpy hubby with the grumbling tummy leaves for work without even a good-bye kiss. "What'd I do to deserve this?" wails the wife, to which Betty has the audacity to reply, "It's what you DIDN'T do. You wouldn't fix the most important meal of the day." She prescribes Shredded Ralston, and at breakfast the next morning, the fuddy-duddy hubby brags, "This is how I dreamed it would be . . . A pretty wife and a good stick-to-the-ribs breakfast!" He should be thankful Betty Bite Size was not selling Wheatena, or his pretty wife might have delivered a *kick* to the ribs instead.

Well, such was the state of cereal characters during the late 1940s. By the time the next decade arrived, the new mass medium of television would be absorbing more and more of the advertising formerly directed through radio, newspapers, and magazines. Some of the familiar figures from the early years would still be around, but television would produce an unprecedented number of new arrivals on the scene. Turn the page, and see how the biggest company and its most legendary advertising agency conspired to help bring this about.

Left: Kellogg's first cereal mascot, the Sweetheart of the Corn, was even more appealing when painted in full color for this 1915 magazine advertisement.

Below: A ninety-second theatrical cartoon, *Breakfast Pals*, was the first animated treatment of Kellogg's three representatives of Rice Krispies, Snap!, Crackle!, and Pop!

TONY THE TIGER, SAYS
"GR-R-REAT!"

NEW from Kellogg's of Battle Creek

The biggest, sweetest thing that ever happened in cereals. Big, crisp flakes—sparkling all over with Kellogg's secret sugar frosting. Really, folks, their flavor is so deliciously different that we doubt whether even the Greeks had a word for it—unless it was GR-R-REAT!

Kellogg's
SUGAR FROSTED FLAKES

Above: When Tony the Tiger was introduced as the logo for Sugar Frosted Flakes in 1952, his design was a bit different from his later, more humanized form. This was one of the first magazine ads for the new product and new character.

Right: Post tried its luck at sugar-coating corn flakes and came up with Corn-Fetti. The mascot character, Captain Jolly the pirate, is often thought to have been the distant ancestor of Quaker's much later Cap'n Crunch.

THE CAPTAIN CLAIMS THE PRIZE . . . *A Captain Jolly Adventure*

I W-WISH WE'D DROPPED THE OLD MAN'S TREASURE CHEST OVERBOARD INSTEAD OF THAT NEW, IMPROVED CORN-FETTI! HE'S SO CRAZY ABOUT!

HE'S MEAN AS A MAN-EATING SHARK WHEN HE DOESN'T GET HIS CORN FLAKES WITH THE MAGIC SUGAR COAT!

BETTER PRACTICE DODGING, COOK! HE'LL CROWN YOU WITH THE SUGAR BOWL IF YOU TRY TO SERVE HIM ANY CEREAL BUT CORN-FETTI. HE SAYS IT HAS REAL MAGIC SWEETNESS!

BLIMEY! THE CAPTAIN'S CRATE OF CORN-FETTI! LET'S PULL HER ABOARD!

ANY TREASURE THAT COMES ABOARD THIS SHIP BELONGS TO THE CAPTAIN! YOU DON'T GET ONE CRUNCHY FLAKE OF THIS CORN-FETTI...OR OF THE OTHER CASE I BROUGHT ALONG!

AH, DIVE IN! I CAN'T STAY MAD WHEN I'M EATING THE CRISPEST CEREAL ON LAND OR SEA! DOESN'T GO MUSHY, EVEN WHEN YOU DROWN IT IN MILK!

NEW KIND OF CORN FLAKES WITH THE MAGIC SUGAR COAT

Now guaranteed not to get sticky, even in opened package, or triple your money back!

Copr. General Foods Corporation   Product of General Foods

Above: **When Kellogg's began sponsoring *Huckleberry Hound* in 1958, it marked the beginning of a long and mutually beneficial relationship between the cereal makers and the Hanna-Barbera cartoon studio.**

Left: **Post's Sugar Crisp was the first nationally distributed, presweetened cereal. Its three bears reminded everyone, "As a cereal it's dandy, for snacks it's so handy, or eat it like candy." Not surprisingly, their names were Dandy, Handy, and Candy.**

Above: This colorful set of children's dishes immortalized most of the Post cereals' character lineup of the era: the three Sugar Crisp bears, plus the Rice Krinkles clown, and the short-lived Grape-Nuts fox.

Right: The Cheerios Kid had been around since 1955, but in the 1960s he was teamed with girlfriend Sue for a memorable series of commercials, and also partnered with Bullwinkle the Moose for another round of slapstick adventures.

Left: General Mills introduced its marshmallow-laden treat Lucky Charms in 1964. Originally Lucky the Leprechaun was a more impish gnome than the benign fellow he became later in the decade.

For Cheerios' comic-book ads such as this one, Bullwinkle appeared with his usual partner, Rocky the Flying Squirrel. In the animated commercials, Rocky's role was filled by the Cheerios Kid.

Above: **The Trix Rabbit began lusting for his favorite fruit-flavored cereal in 1959, and he continues his usually fruitless quest more than fifty years later.**

Right: **Imitating Big Otis the Scotsman, Yogi Bear flexed a muscle on the packages of Kellogg's OK's. In the single-serving assortment, he shared shelf space with Crackle!, Corny the Rooster, and Sugar Pops Pete.**

Left: When Kellogg's introduced Froot Loops in 1963, the original design of mascot Toucan Sam had a beak stretching most of the width of the box and a tall hat laden with tropical fruit.

Below: While so many rural-themed sitcoms (nicknamed "corncoms") were filling up network TV airtime in the mid-1960s, Kellogg's joined the fun by assigning the omnivorous Hillbilly Goat to sell Sugar Stars.

Right: In 1964, Post took the bold step of turning its gang of cereal mascots into a half-hour Saturday-morning series, *Linus the Lionhearted*. This record album is one of the most sought-after relics from the period, featuring the voices of Sheldon Leonard (Linus), Carl Reiner (Billy Bird), Gerry Matthews (Sugar Bear), Ruth Buzzi (Granny Goodwitch), Jesse White (Claudius Crow), and Robert McFadden (as everyone else).

Below: Tony the Tiger had sired quite a large family by the time of this 1968 newspaper ad. Notice the illustrations of the current Kellogg's packaging, featuring the relatively new arrivals the Smackin' Brothers (for Sugar Smacks), the Whippersnapper (for Sugar Pops), Apple Jack, and Ogg the Caveman (for Cocoa Krispies).

Left: **This metal lunch box was available in 1969. It pictured most of the Kellogg's cast, but like the single-serving boxes, some of these character designs were already obsolete when the lunch box hit the market.**

Below: **The single-serving sizes of the Kellogg's boxes became miniature time capsules. Several of the designs seen in this late-1960s ad were still in use in the mid-1970s, long after the characters had been either revamped or eliminated on the regular-size packaging.**

Above: Lovable Truly the Postman (*not* a mailman!) delivered the "letters" of Post Alpha-Bits from 1964 until the early 1970s.

Right: Post introduced Fruity Pebbles in 1969, followed closely by Cocoa Pebbles the next year. The clever commercials delivered by the Flintstones made them a favorite target of those who wanted more separation between program content and advertising in children's TV.

Above: Nabisco's Wheat and Rice Honeys were advertised by Buffalo Bee from the mid-1950s to the mid-1960s. Premiums tended toward a western theme, as illustrated by this ad from a Buffalo Bee comic book.

Left: Cap'n Crunch was one of the most successful cereal icons ever, and it was the first time a product and its mascot character were created together and shared the same name. Occasional non-premium items were developed for retail stores, such as this coloring book from 1968.

Right: **These single-serving boxes from 1975 were the last appearances of the fruit-headed Apple Jack (who had vanished from the full-size boxes in 1968–69) and the original long-beaked version of Toucan Sam (who had been redesigned in 1969).**

Below: **General Mills introduced its first two "Monster Cereals," Count Chocula and Franken Berry, in 1971. The earliest commercials gently spoofed classic horror-movie clichés, but the two creepy creatures most often succeeded in frightening each other.**

Above: **In the mid-1970s, the appearance of Snap!, Crackle!, and Pop! had come a long way from their roots in the early 1930s. By this time, their voices in the commercials were being supplied by busy cartoon actors Daws Butler and Don Messick.**

Left: **Just as the 1968 ad we saw earlier illustrated the Kellogg's packaging of its era, so does this 1979 ad with a back-to-school theme. New arrivals during the interim included Dig'em the Frog, Big Yella, and Tusk Tusk the Elephant.**

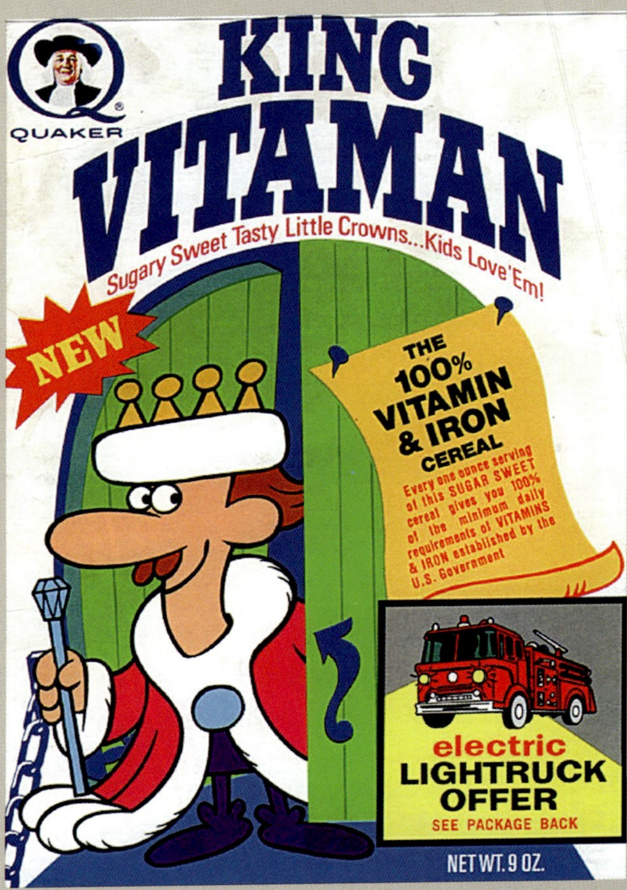

Right: After his amazing Quaker ad campaigns for Cap'n Crunch, Quisp, and Quake, Jay Ward helped the company once again in 1970 with the early commercials for King Vitaman.

Below: If frosted corn flakes were going to be colored pink and given a strawberry flavor, who would have been a more fitting mascot than that silent cool cat, the Pink Panther?

Above: In the mid-1980s, General Mills cosponsored a traveling marionette show with a crime prevention theme. The Trix Rabbit, Sonny the Cuckoo, the Honey Nut Bee, Count Chocula, Franken Berry, and Lucky the Leprechaun all joined McGruff the Crime Dog for important lessons about safety.

Left: Ralston marketed innumerable cereals during the 1970s, most of which tasted the same and lasted about a year. The most memorable of all was Freakies, due to its incessant animated commercials and the sheer weirdness of its seven namesake characters.

Above: Quaker Oats did not consider the ramifications when they enlisted Popeye as the mascot for their instant oatmeal. When he proclaimed, "I'm Popeye the Quaker Man," however, the famously pacifist Quaker church reacted with some not-so-peaceful outrage.

Right: This was one of the last boxes to feature the original design of Cap'n Crunch. After Jay Ward gave up the ad campaign, and the Cap'n's voice, Daws Butler, died in 1988, the old salt was given a younger look and a less doddering voice.

The Best to You Each Morning

Kellogg's in the 1950s

When it became evident that video was going to be a major outlet for advertising, Kellogg's reacted by firing its established New York ad agency and throwing its business to the Leo Burnett agency of Chicago. Shades of the Sunny Jim tale, Leo Burnett and his agency had long been sneered at by the more elite in the advertising business. Burnett had a grassroots, down-to-earth approach to advertising that seemed positively homely to his competitors, yet his work with Kellogg's and countless subsequent clients would grow his company into an advertising powerhouse whose megaton impact the general public would not even recognize.

Just before Burnett geared up for new battles on behalf of Battle Creek, Kellogg's previous agency had launched the company's television career with two series, a video version of Ireene Wicker's *Singing Story Lady* radio show and a space opera, *Tom Corbett, Space Cadet*. The latter was the more successful of the pair, chronicling the spaced-out adventures of the title character, played by former movie actor Frankie Thomas. The show was set four centuries in the future, in AD 2350 to be precise, and depicted the space academy's efforts to maintain peace throughout the universe.

Periodically, the characters would have to pause in their strenuous efforts long enough to sit down and have a hearty Kellogg's breakfast. In retrospect, it seems somewhat amazing that in four hundred years, not only had the cereals remained the same, but their packaging still looked exactly as it had in 1950.

Kellogg's offered a number of *Tom Corbett* premiums but generally did not splash the characters' pictures all over the fronts of the boxes, at least not to the extent that competitor Ralston did with its own Corbett TV clone, *Space Patrol*. In fact, Kellogg's packaging of 1950–51 was about the blandest of any cereal company. Most of the boxes were no more than plain white with the product's name in huge black letters and only the red Kellogg's logo to break up the monotony. The three faces of Snap!, Crackle!, and Pop!, each wearing his distinctive headgear, were added to the Rice Krispies boxes almost as an afterthought.

Pep, the cereal that had been around since Superman's radio days, spiffed up its box with a logo that had even less personality than the Cream of Wheat chef. In fact, the chef and the Quaker Oats man had the advantage over Pep's new "character" in that at least they had heads. The Pep package featured artwork of a brawny arm—presumably male—flexing a bulging bicep. With such a prominent display of sinew, one would have expected the commercials to fall into the "instant strength" ploy of the Popeye/Wheatena spots, but in fact, the arm seemed to have very little connection at all. Each television spot would dissolve from the artwork of the arm to a live actor with his or her appendage in roughly the same position, but *not* flexing a muscle. As the performers mimed their routines, the announcer would go on and on about Pep's ability to prepare kids for their future work of being high school football players (for the boys) or energetic—dare we say peppy?—cheerleaders (for the girls).

It is likely that Leo Burnett and his like-minded staff simply cringed at the blandness of the Pep arm as a logo. When given a chance to sink his choppers into a heaping-size helping of Kellogg's newest product, though, Burnett proved things were going to be different around good old Battle Creek in more ways than one.

For years, Kellogg's representatives had noticed people "improving" on the natural flavor of even its best-selling products by dousing them with sugar before consuming them. True, this did add a certain kick to Corn Flakes and Rice Krispies, but when it came time to take the natural plunge and make a cereal that was already sweetened, the Kellogg heritage in the Seventh-Day Adventist sanitarium caused the company some true soul-searching. Having begun as a health food, some felt going into the sugar-coated cereal business would be a mortal sin, but just as W. K. Kellogg had won out over brother John Harvey when it came to marketing their Corn

For a brief time in the early 1950s, Corn Flakes revived the appealing visage of the Sweetheart of the Corn. She had, however, been modernized to look like a sort of cross between Dorothy of Oz and Rebecca of Sunnybrook Farm.

Flakes in the first place, the company finally shut its eyes and prayed it was not venturing down the wide path to destruction.

The product that resulted, in early 1951, was originally called Sugar Corn Pops but just as often went by an abbreviated moniker, Sugar Pops, or sometimes Corn Pops. This was to be Kellogg's first product launched primarily through television, and the Leo Burnett agency spared no effort in proving just where its expertise lay.

Kellogg's paid for the development of a children's western concerning the fictitious adventures of the real-life lawman "Wild Bill" Hickok. On the tube, the role of Hickok would be essayed by Guy Madison, but at least as much attention was given to his portly comic-relief sidekick, Jingles. Yes, it was our old friend Andy Devine in his latest role, showing he could chew up scenery as efficiently as he could chow down a meal at the dinner table. The drab old black-and-white Kellogg's packaging was relegated to the scrap pile, and both Madison and Devine appeared in full color on the Sugar Corn Pops box fronts, which were now the loudest shade of yellow that could possibly be run through a printing press. This effect was lost on the black-and-white TV screen, but certainly made up for that deficiency by leaping out at adults and kids alike from the grocery store shelves.

The commercials' jingle—although it was not sung by Jingles—emphasized that the product went under two names. "Kellogg's Sugar Corn Pops . . . (BANG! BANG!) Sugar Pops are tops!" Madison and Devine might have been the first characters to use their own commercials to comment on their pictures appearing on the cereal boxes. In these amusing spots, they were in costume but obviously portraying themselves as actors rather than the historical figures of their show. One commercial begins with a boy and girl sharing a box of Sugar Pops with Guy/Wild Bill on the front, as Devine eavesdrops on their conversation:

BOY: Gee, isn't this a swell picture on the front of the box?
GIRL: I'll say! He sure is handsome, isn't he?
BOY: Yeah, he looks real rugged, too!
ANDY: Aw, shucks, you shouldn't say those things in front o' me . . . You'll make me self-conscious!
GIRL: Oh gee, Andy, we were talking about Guy's picture!
ANDY: Aw, Guy ain't so handsome.
GUY: (coming up behind Andy) What's that?
ANDY: I said . . . (double take) . . . oh (giggle) . . . howdy, handsome!

GIRL: Gee, Jimmy, we better pass these Sugar Corn Pops before Andy REALLY gets in trouble.

GUY: Now you're talking! Who cares what's on the box as long as you get what's in it?

ANDY: That's right! Get a box of Kellogg's Sugar Corn Pops, kids! Guy and I know you'll really go for 'em!

About a year after Sugar Pops popped out, Kellogg's took an even bolder step and, over the objections of some of the company old-timers, found a way to add a layer of sugar to reliable old Corn Flakes. Sugar Frosted Flakes were an immediate hit, both for their taste and the innovative advertising program devised by the Burnett agency.

The animated commercials for Frosted Flakes were built around the unlikely theme of jungle beasts—or at least they were seen living in the jungle, regardless of what environment their real-life counterparts existed. Katy the Kangaroo was usually seen with her young in her pouch and in overall personality and appearance reminded one strongly of Winnie-the-Pooh's companions Kanga and Roo. The identity of the lady who supplied Katy's voice remains a mystery.

There is certainly no mystery about the voice given to Frosted Flakes' other spokes-character. In fact, Tony the Tiger was the first Kellogg's trademark to be so identified with a distinctive voice. Thurl Ravenscroft was a singer and actor who, in 1937, had formed the male quartet known as the Sportsmen. After World War II, since the Sportsmen had found a new bass singer while he was serving overseas, Ravenscroft formed a new quartet called the MelloMen. By the time Frosted Flakes were rolled out, the MelloMen were being heard regularly in commercials, in animated Disney features, and in TV-show theme songs (such as the *Hopalong Cassidy* theme). Somehow, Ravenscroft was separated from his comrades to give voice to Tony, a role he would play for the next fifty-three years.

In the interest of historical accuracy, it must be noted that another well-known cartoon voice actor, Dallas McKennon, claimed to have been the original voice of Tony. According to his version of things, the growl he supplied was too much of a strain on his vocal chords, requiring Ravenscroft's naturally deep tones to be brought in. To date, no early commercials with McKennon as Tony have been found, but his story is given credence by the fact that McKennon can indeed be heard as one of the supporting jungle animals, George the Giraffe, in some of the early spots.

Like the question of his voice, there is also some dispute when it comes to Tony's graphic design. Many sources insist the original box art was the

The biggest, sweetest thing that ever happened in cereals. Every crisp flake is sparkling all over with Kellogg's secret sugar frosting. Honest, folks . . . one taste is worth ten thousand words.

When Sugar Frosted Flakes debuted in 1952, some of the boxes featured Katy the Kangaroo, and others featured Tony the Tiger. After a year or so, Katy hopped back into obscurity and left Tony as the undisputed king of the Kellogg's jungle.

work of famed children's book illustrator and former Disney artist Martin Provensen. However, an equal number of sources counter that Tony's appearance was not Provensen's work, but was merely an imitation of his style. In either case, the box-front Tony bore only the slightest resemblance to his present-day look. His head was shaped like a football, and he had paws instead of humanized hands and fingers. Two elements that have been retained for tradition's sake, even as Tony's look was streamlined over the years, are his blue nose and the red kerchief around his neck.

Regardless of who designed the characters, for approximately a year, Frosted Flakes boxes and advertisements featured Tony and Katy equally. By late 1953, it was obvious that Tony had the most personality, so Katy was quietly phased out and forgotten. In print ads, Tony was not particularly anthropomorphic; instead, he looked and acted more like a real tiger—albeit a grinning, cheerful one. (One ad parodied the famed "I'd walk a mile for a Camel" cigarette spots by having Tony declare, "I'd stalk a mile for Kellogg's Sugar Frosted Flakes.") He was frequently accompanied by a young tiger cub, presumably his offspring, whose role would become more important during the 1960s and 1970s.

It seems there should be some obvious answer as to just why Tony was such a breakout star, while Katy crawled back into her own pouch and disappeared, but no single reason seems to stand out above any other. Let's face it: tigers have always been more popular in pop culture than kangaroos, so maybe it was a simple matter of species preference. Ravenscroft's voice certainly brought with it a warmth and humor that were outgrowths of the performer's real-life personality, so perhaps it was that vocal quality that made kids want to listen to him. As stated in this book's preface, the most successful cereal mascots have been those who were timeless, while the ones who tried to be hip or take advantage of current trends were usually short-lived. Tony might have been the most timeless of all; other than losing some of his early features such as claws and fangs, he has managed to survive for more than six decades to become undisputed king of the cereal aisle.

Many of Tony's TV appearances during his early period were tied to Kellogg's sponsorship of the new children's show *The Adventures of Superman*, starring George Reeves. Several spots began with Tony's image on the front of the cereal box coming to life and proclaiming, "I want you to meet a friend of mine! Clark Kent, star of *Superman*!" (Apparently Sugar Pops' method of identifying actors by their real names was not wanted in this series.) Then Reeves, as Clark, would take over, sometimes using

In Tony the Tiger's first appearances, in both print and animated form, he was presented as a much more realistic tiger than the humanized form he possesses today. This early magazine ad even depicts him walking on all fours in a "stalking" pose that would be quite foreign to the Tony of the 1960s and 1970s.

his X-ray vision to peer through walls and see how much the people inside were enjoying Frosted Flakes. Instead of being arrested as a peeping Tom, Reeves/Kent would end by asking Tony what he thought of Frosted Flakes. This was the key for Ravenscroft to deliver one of the most famous slogans in all advertising history: "Gr-r-r-r-reat!"

Along with Tony's ever-increasing appearances, the Burnett agency also found new ways to pep up those old veteran gnomes Snap!, Crackle!, and Pop! On the box fronts they received scrawny new bodies, with their oversized heads rather out of proportion, but in the animated commercials they hardly resembled the artwork at all. After years of Snap!'s chef outfit and Pop!'s military attire, Crackle! finally received a new outfit to go along with his stocking cap and was seen wearing what appeared to be a clown suit. Shades of Kellogg's sponsorship of *Tom Corbett*, their new commercials depicted them as residents of faraway Planet K, zooming to earth in a flying saucer, with a flying bowl to go along with it. At this point, they were still not given any sort of meaningful dialogue, but the singing in the commercials—as well as the elves' occasional words—was supplied by the MelloMen.

In recent years, a story has circulated among nostalgia fans that a fourth member of the Krispies team was added during this period. So far, the only evidence of this fourth wheel, Pow!, is a piece of concept art depicting him as a brawny superhero or spaceman type. It is very possible that he never made it from the drawing board to either the cereal box or the small screen; if he ever did, it must have been only a brief time before Pow! got it right in the kisser.

While Tony the Tiger was growling his approval of Frosted Flakes and Wild Bill and Jingles continued shopping for Sugar Pops, Kellogg's brought out its next presweetened treat, Sugar Smacks. In a rare example of following rather than leading, Kellogg's somewhat sticky snack was indistinguishable from two other products already on the market from Nabisco and Post. Sugar Smacks could have desperately used an identifying character as strong as Tony, but in the beginning, Burnett apparently decided to take a different route.

The Sugar Smacks boxes were dominated by the giant faces of two of the Ringling Brothers Circus's most famous clowns, Lou Jacobs and Paul Jung. There was no apparent connection between the circus theme and any inherent characteristic of Sugar Smacks. Possibly the association came about because Kellogg's was a partial sponsor of the kids' TV show *Super Circus*, but if that were indeed the case, the connection was tenuous at best. Jacobs and Jung not only loomed large on the boxes but were

The Rice Krispies boxes of the early 1950s gave Snap!, Crackle!, and Pop! some scrawny bodies quite out of proportion to their oversized heads.

featured in the giveaway premiums too, just as if they were Sugar Smacks' logo. (It is tempting to wonder how this ad campaign affected children who happened to have an inborn fear of clowns—but then again, in the 1950s clowns had not developed the sinister subtext that movies and novels of the 1960s and 1970s would impart to them.)

Perhaps for this reason, but more likely because of the cost of paying for the live clowns' endorsement, in 1957 they were sent back to the circus and a new Sugar Smacks spokes-mammal was introduced: Smaxey the Seal. This maintained some continuity with the established circus theme, but somehow poor old Smaxey failed to make anything near the same impact as Tony the Tiger. In his commercials, Smaxey did not even have a voice, much less a distinctive one. While barking "arf, arf," the only thing Smaxey did to justify his connection with Sugar Smacks was the slapping of his flippers while he balanced a bowl on his nose.

Meanwhile, the immigrants from Planet K—Snap!, Crackle!, and Pop!—were undergoing some changes of their own. By 1958, they had ceased to appear as a group on the Rice Krispies boxes. Instead, somewhat like the clown faces on Sugar Smacks, only one elf's giant head would be featured on each box. (The magazine ads pointed out this fact and reassured consumers that the other two could be found on additional boxes.) This trial separation of the trio applied only to the box artwork, as their television commercials continued to present them as a unified entity.

It was at this same time that Kellogg's took on sponsorship of the *Woody Woodpecker Show*, a weekly afternoon compilation of the classic Walter Lantz "Cartunes," as he had trademarked them. The Lantz studio was enlisted to produce special Kellogg's commercials featuring Woody to be aired during the show, and for these, the writers pulled out the old "instant strength" chestnut of an idea that dated back to Popeye and Wheatena. So what if Woody Woodpecker had never been presented as a specimen of physical fitness?

In one of the best of these commercials, loaded with the rapid-fire sight gags that were part of the Woody cartoons, the pointy-beaked bird is seen merrily rushing to the supermarket with Rice Krispies at the top of his shopping list. Once inside, his good mood is dampened when he sees a cigar-chomping bully, who could be Bluto's beardless brother, greedily cleaning out every single box of Rice Krispies from the shelf. Feisty as usual, Woody tries to stop the palooka with physical force, but the sneering miscreant picks him up by the topknot and tosses him into a nearby shopping cart.

From 1958 until the early 1960s, Snap!, Crackle!, and Pop! appeared on the Rice Krispies packaging one at a time, the only instance of this inseparable trio being split up.

Although he was never before presented as a shining specimen of physical fitness, Woody Woodpecker gained instant superstrength from eating Rice Krispies, thanks to the timely intervention of Snap!, Crackle!, and Pop!

With Woody aboard, the cart careens through the store, wrecking the displays. (In one great moment, it crashes through a pyramid of canned goods, causing them to rearrange themselves into the word HELP.) From a nearby sign, the printed images of Snap!, Crackle!, and Pop! finally catch on to what is happening. "It's Woody! He needs help!" rasps Crackle! (another Dallas McKennon voice). The elfin trio finds a dazed Woody on the floor, and they quickly break out a box of Rice Krispies, a bottle of milk, and a bag of sugar. While Woody chomps on his meal, the jingle points out the cereal's three main selling points: "Snaps with energy, crackles with fun/Pops up the muscles on everyone!" Sure enough, Woody flexes a huge bicep that only momentarily fazes the bully. Thinking quickly, the brute extends a hand in phony friendship, only to have Woody grab it in an iron vice grip. For the ending, the bully watches the crushed bones in his injured hand go snap, crackle, and pop.

In another adventure from the same era, Woody heads into the woods, wearing a hunting cap and toting a rifle, loaded for bear. Unfortunately, the bear is also loaded for him and chases Woody off a cliff where Snap!, Crackle!, and Pop! are waiting. With no grocery store nearby, they rush into a neighboring Kellogg's factory (maybe this forest is near Battle

Sugar Pops Pete was introduced in 1958. At no other time was there ever a Kellogg's mascot character whose appearances on the box and in animated form differed so greatly. Pete used his "pop gun" to sweeten the disposition of the West's most ornery outlaws.

Creek) and prepare the necessary bowl of strength for their pal. After eating the cereal, Woody picks up the bear and tosses him into the distance.

Sugar Pops were still doing fine with Guy Madison and Andy Devine serving as identifiable characters on the boxes and in the commercials, but in 1958, the inevitable happened. *The Adventures of Wild Bill Hickok* was finally canceled, and Kellogg's had to go looking for someone new to convince youngsters that Sugar Pops were tops. Sticking with a proven formula, Sugar Pops would continue their western setting but with a new character who was not tied to any particular series: Sugar Pops Pete.

Pete was a prairie dog who dressed like a miniature Wild Bill. It must be admitted that in no other instance, past or future, was there ever a Kellogg's logo character whose appearance on the box and look in the animated commercials differed so vastly. In fact, one would be hard-pressed to identify the printed Pete and the animated Pete as the same character.

Pete's commercials were spoofs of the ubiquitous TV westerns of the period. Wielding his candy-striped "pop gun," Pete explained his modus operandi in a theme song that recounted how his magical gun could turn the most despicable bad guy "sweet." Pete's fructose-enhanced rehabilitation of ornery outlaws continued into the mid-1960s.

Another new product was unveiled in 1958. Taking the same principle that had converted Corn Flakes into Frosted Flakes, Kellogg's found a way to add a bittersweet chocolate flavor to Rice Krispies and introduced them to the world as Cocoa Krispies. The first character cast to advertise the new product, both on the boxes and in the commercials, was a monkey in a straw hat and tropical shirt. Most sources identify this simian as José, but there are also researchers who have referred to him as Coco. Regardless of what he was called, the monkey only monkeyed around with Cocoa Krispies for a brief time before being sent back to his tree. His catchphrase outlasted him, to be delivered by other characters in the future: "Tastes like a chocolate milkshake, only crunchy!"

Kellogg's biggest move into television yet began in October 1958. The newly established Hanna-Barbera cartoon studio had made its debut the previous year with a series on NBC, *Ruff and Reddy*, but it was Kellogg's backing that enabled the studio to introduce its breakout hit of the era, *Huckleberry Hound*. Unlike *Ruff and Reddy*, *Huckleberry Hound* was not seen on a network, nor was it a Saturday-morning show. Instead, Huck was part of a rotating group of shows (including *Woody Woodpecker* and *The Adventures of Superman*) Kellogg's sponsored in syndication, a different series airing in the same time slot each weekday afternoon. Since

The original opening title sequence for Kellogg's *Huckleberry Hound* series featured the show's star only in this brief cameo. The rest of the theme song was acted out by Corn Flakes' new mascot, Cornelius the Rooster.

The rest of the 1958 Kellogg's cast joined the fun during the closing credits for *Huckleberry Hound*. The easygoing mutt and Cornelius the Rooster gathered Snap!, Crackle!, Pop!, Tony the Tiger, Tony Jr., Smaxey the Seal, and Sugar Pops Pete for a rousing finale.

Huckleberry Hound consisted of all-new animated material, it was the one that attracted the largest fan base.

There was no doubt as to just whose money was paying for this show. The opening titles introduced a new Kellogg's character—and it was *not* Huckleberry Hound. The first shot on the screen was of a newly redesigned Corn Flakes box featuring the brand's first true logo character since the days of the Sweetheart of the Corn: a somewhat abstract representation of a green, red, and yellow rooster, meant to connect the product with Kellogg's slogan, "The best to you each morning." With a rousing "cock-a-doodle-doo," the logo rooster sprang to life as a fully rounded, animated character who cavorted through the rest of the theme song. This was all quite unusual since other than a quick glimpse of the show's title, Huckleberry Hound himself never even appeared in the opening of his own show. Instead, the rooster—who would eventually be given the appropriate name Cornelius, or Corny for short—danced his way through a circus setting, playing a ring-the-bell game and joining an elephant on a teeter-totter.

Once the opening was finished, Cornelius would make one more appearance, knocking on Huck's dressing-room door, before disappearing to turn the rest of the show over to the stars. Besides Huck's easygoing adventures, there were additional segments with the mice Pixie and Dixie and their feline nemesis Mr. Jinks ("I hate meeces to pieces!"), and that smarter than average plantigrade, Yogi Bear.

Anyone viewing the commercial breaks during a complete half hour of the *Huckleberry Hound* show in 1958 would have seen any or all of the other Kellogg's characters we have discussed. Just to drive home the sponsor's message, though, they all appeared en masse for the closing credits. Cornelius the Rooster got top billing once again, driving a jalopy around the circus's three rings with Huck Hound hanging on for dear life. This particular circus seemed to have an amazing and motley assortment of hitchhikers in the sawdust, because in rapid succession, the chicken-driven auto would pick up Snap!, Crackle!, and Pop! (Crackle! still wearing his clown suit from the commercials earlier in the decade), and, in a group, Tony the Tiger and his son, Smaxey the Seal, and Sugar Pops Pete. As the credits ended, Tony Jr. struck a hammy pose on top of Huck's ringmaster hat, only to be knocked off by the car exiting through the small gap forming the circus-tent entrance.

The immediate success of *Huckleberry Hound* made Kellogg's antsy to get another Hanna-Barbera show on the air, and the result was 1959's *Quick Draw McGraw*. This time, the Kellogg's mascots did not intrude upon the show's standard cast of characters nearly as much, but the program found a hungry audience waiting to gulp it down. There was a similar reaction when Kellogg's loaned its sponsorship to the live-action *Dennis the Menace* series, starring towheaded Jay North as a somewhat older version of Hank Ketcham's famous comic-strip character.

These series served as a platform to introduce what would be Kellogg's last new cereal, and last new ad character, for the decade. It must have been grating on the Kellogg executives' nerves to find, while their cereals paralleled most others on the market, the company had never managed to come up with its own competitor to General Mills' Cheerios. Thanks to some clever animated commercials originating in the wonderful world of Disney, Cheerios were enjoying even greater success than before, and Kellogg's was left on the bench to watch.

Big Otis, a brawny Scotsman, displayed his might on Kellogg's new product, OK's, in 1959. Even the company's executives found his presence irritating, and his role as box-front mascot was soon taken by Yogi Bear.

Not for long. In 1959, Kellogg's okayed OK's, an oat cereal that went Cheerios one letter better by adding a K (for Kellogg's, naturally). For the first commercials, Leo Burnett dragged out the old association of oats with Scotland—as the first marketers of oatmeal had dealt with a century earlier—and introduced Big Otis, a bearded giant of a Scotsman wearing kilts and a sleeveless shirt to display his massive arms. The Otis of the box front was artwork, but a live actor—if his limited activities could be called acting—portrayed him in the commercials.

As Big Otis stamped down the bonnie trail with young children trailing him, the announcer tried to affect a proper Scottish burr, touting OK's as "made forrr brrrawny lads and bonnie lassies." At the end of each spot, the unseen narrator would get rolling again by reminding everyone OK's were "a bonnie brrreakfast." Big Otis had little to do but look strong and lift kids into the air onto his shoulders. Veterans at Leo Burnett told cereal historian Scott Bruce that Big Otis was short-lived because Kellogg's president, Lyle Rolls, personally hated him. The exact reasons for Rolls's dislike of Big Otis are not clear, but the burly brute would be reunited with his Bonnie over the sea by the early 1960s.

Now, never let it be said that Kellogg's and Burnett had some sort of monopoly on characters, both successful and otherwise, during the 1950s. In our next chapter, we shall take a look at how Post was determined not to be left at the post in the great cereal mascot derby.

For Breakfast It's Dandy

Post in the 1950s

Even before Kellogg's decided to take the plunge into the presweetened cereal market, archrival Post—which, as we have seen, developed out of the same health-food background in Battle Creek—had been keeping a weather eye on Ranger Joe and became convinced sugar was the way to go, Joe.

Now, who or what was Ranger Joe? That was the name given to the first presweetened cereal in history, developed in Pennsylvania in 1939. Ranger Joe was basically puffed wheat with a sugar glaze coating each puff. The product was heavily promoted—even today Ranger Joe cereal bowls and milk mugs are not too uncommon in antique stores—but the inescapable fact was that Ranger Joe's range was rather limited. Outside of Pennsylvania and the adjoining states, Ranger Joe was a stranger in them thar parts, and it was Post's contention that such a presweetened product could have national appeal.

Did they put their kitchen staff to work to come up with a new type of cereal? No, they simply took the basic sugar-coated wheat-puff idea and manufactured it themselves, calling their new product (well, at least it was new to non–Keystone State residents) Sugar Crisp. It was sprung on an

In late 1949 and early 1950, the three bears named Dandy, Handy, and Candy helped introduce Post's new presweetened cereal Sugar Crisp to the general public. Notice, in this early ad, that Sugar Crisp still came in cellophane bags instead of boxes.

unsuspecting public in 1949, and by the time 1950 arrived, its advertising was occupying full pages in national magazines.

Since most of the future Kellogg's celebrities were still a few years away from their debuts, when Post decided Sugar Crisp needed some mascots to help sell the product, the Benton and Bowles ad agency apparently turned to Snap!, Crackle!, and Pop! the same way the developers had spied on Ranger Joe. Sugar Crisp's first ad characters would be three identical bears, given the names of Dandy, Handy, and Candy.

This trio of monikers was meant to remind consumers of the three main selling points of Sugar Crisp, which were hammered into every magazine and newspaper ad: "As a cereal it's dandy; for snacks it's so handy; or eat it like candy!" The choice of bears was a little more esoteric but likely had something to do with the wheat puffs' honey-flavored coating. Occasionally they would even be referred to as the "honey bears," but more often simply as the "Sugar Crisp bears."

The furry threesome began appearing in animated television commercials almost as soon as the product appeared on the market, with their major appearances during two Post-sponsored westerns, *Hopalong Cassidy* (1950) and *The Roy Rogers Show* (1951). Over the approximately seven years represented by those two combined sponsorships, the Sugar Crisp bears' commercials adopted perhaps the widest variety of formats of any single cereal's ad campaign.

Most of the commercials were done in what would come to be recognized as standard television animation: not as lavish as the Disney, Warner Bros., or MGM theatrical cartoons, but better than static comic-book panels. Occasionally there would be radical departures from this norm. There are existing examples of the bears, for example, rendered as stop-motion animated puppets, an expensive and time-consuming form of animation not often employed for children's commercials. (It was slightly more common for advertising adult-oriented products such as beer, cigarettes, and Alka-Seltzer.) Even more strangely, at least a few commercials did away with the animated bears completely and used footage of three live black bear cubs at play, while a narrator described what they were supposed to be doing and thinking.

Even when Dandy, Handy, and Candy were themselves, one aspect of their spots that just could not seem to remain consistent from appearance to appearance was their voices. On the occasion when one of them spoke alone, he (or she, since their sex was never specified) would frequently sound like a young child—but not always. Other commercials featured musical narration by a male quartet that sounded like vocalist Ken

Stop-motion animation was usually too expensive and time-consuming to be practical for children's television commercials, but at least one exception brought the three Sugar Crisp bears to life using that medium.

Darby's group known as the King's Men; three of the four singers would be employed to speak individually for the bears' minimal dialogue. Still another run of commercials had a sound track by another vocal group, giving one bear a deep-bass, Tony the Tiger–type voice and another bear a distinctly female one. (We told you their gender was ambiguous at best.)

No matter how they sounded, the three bears were more often than not seen using a bowlful of their sweetened wheat puffs to stop some sort of trouble in its tracks by distracting the villain, whether it be a shootout in an old western town or a beautiful princess imprisoned by a wicked giant. One thing the bears did *not* do was gain superhuman strength by eating the cereal—but there's no need to fear, Super Sugar Crisp will eventually be here . . . just not right now.

Sugar Crisp was so successful that Post did not waste any time coming up with another presweetened treat to stand alongside it on grocers' shelves. The company took the basic formula of Rice Krispies—no doubt when Kellogg's was not looking—and coated the rice globules with a sugar flavoring similar to Sugar Crisp's but without the honey. The new product's name was tweaked more than once, but soon ended up as Sugar Rice Krinkles (and later, simply Rice Krinkles). The initial mascot character did

For the BEST Choice—
Make your FIRST Choice POST-TENS!

WE LIKE A CHANGE OF CEREAL...

WE HAVE OUR CHOICE OF SEVEN,

AND EVERY DAY WE TAKE OUR PICK—

POST-TENS IS BREAKFAST HEAVEN!

Here's a tip for budget-watchers! Each Post-Tens package holds just one serving, so there's never any waste . . . nothing to get stale or soggy on the kitchen shelf!
Remember—for the *best* choice, make your *first* choice Post-Tens!

GOOD NEWS!
The only assortment with TWO leading sweet-coated cereals—
SUGAR CRISP ...and KRINKLES!

Post's RAISIN BRAN
Post's GRAPE-NUTS FLAKES
Post Toasties CORN FLAKES
Heap Good!
Post's Krinkles
NEW! CANDY-KISSED RICE
Post's SUGAR CRISP
Candy-Coated Puffed Wheat
Post TENS

10 Individual Packages
7 Delicious Health-Builders to Choose from!
3 Post Toasties
2 Grape-Nuts Flakes
1 40% Bran Flakes
1 Grape-Nuts
1 Raisin Bran
1 Sugar Crisp
1 Krinkles

A Product of General Foods

Krinkles—the new rice cereal that's sugar-coated!

...and SUGAR CRISP—as a cereal it's dandy—or eat it like candy!

Freedom needs Your vote!
Vote November 4th

This charming magazine ad from 1952 featured not only a typical family of the era and the Sugar Crisp bears, but also Post's newest ad creation, a piece of the new Sugar Rice Krinkles cereal wearing a clown suit.

The Krinkles clown appeared full-figure on the front of the boxes beginning in 1955 and in live action during the commercials. Post had to hope the vast majority of its kiddie consumer base was not overly afraid of clowns.

not have a lot going for him . . . her . . . it, including a name. The creature was merely a piece of the cereal, wearing a clown suit.

By 1955, the Krinkles Klown (as we might call him) had been replaced by realistic artwork of a human clown, whose image filled the front of the box as did the Ringling Brothers' jesters on Kellogg's Sugar Smacks—which were, of course, virtually the same product as Sugar Crisp—at roughly the same time. Historian Scott Bruce referred to these Krinkles boxes as "sinister," but that probably depended on one's personal attitude toward clowns. Regardless of whether the clowns were seen as benign or threatening, Krinkles would have to wait another few years to gain a more distinctive mascot.

Speaking of distinctive figures, at the time all of this was happening there were few figures as immediately identifiable as William Boyd in character as Hopalong Cassidy. He was set apart from all the many other cowboy stars of the era by his all-black costume—even though he was one of the *good* guys—and his silver hair, which separated him from the usual, more youthful cowpoke crowd. Because of this recognizability, Boyd/Hopalong was used for a time as the front-of-the-box logo for Post

Toasties, Grape-Nuts, and Raisin Bran. Cereal companies were usually loath to lean too heavily on real-life personalities as their mascots because they were more expensive and less dependable than fictional ones, so within a year or two, Hoppy hopped away and the cartoon parade resumed in earnest.

Toasties needed all the help they could get to stand out from the crowd, since, for all practical purposes, they were the same as Kellogg's Corn Flakes. While Kellogg's was making goo-goo eyes at shoppers with a revival of the Sweetheart of the Corn, Post went the comical route and touted Toasties as "the best thing that's happened to corn since the Indians discovered it." Naturally, this was the cue to drag out some Native American stereotypes and splatter them all over the boxes and advertising like war paint.

While such ethnic caricatures seem insensitive today, in 1951 they were well-accepted parts of pop culture. Any negativity they generated could be chalked up to ignorance on the ad agency's part, rather than maliciousness. Still, it makes one toss their Toasties a bit to see ads with

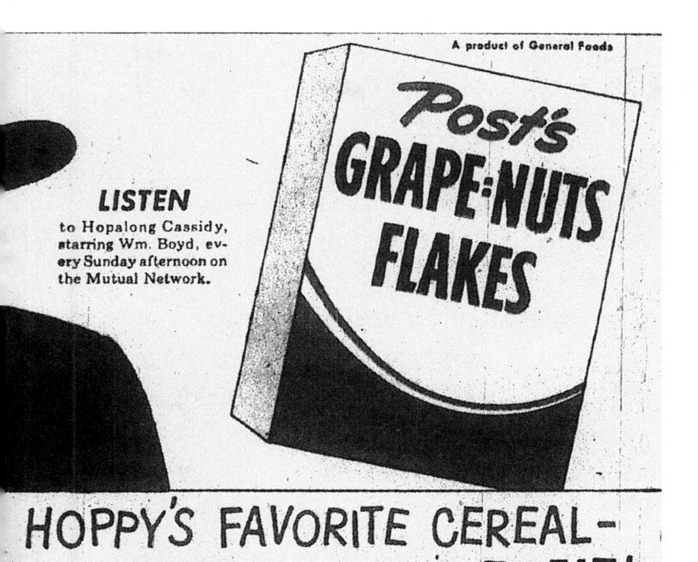

As sponsor of the *Hopalong Cassidy* TV series, Post Grape-Nuts was happy to use the silver-haired, black-suited cowpoke as its main logo character.

befeathered kids spouting platitudes such as "Me wantum stayum-fresh Post Toasties . . . No mush up in milk" and "Me big muscle man, me eatum Post Toasties." The box front featured a single juvenile brave, accompanied by the slogan, "Heap good corn flakes!"

Even with this step backward—at least in some people's view—Post also made a step forward during this period. Prior to 1951, the Post boxes were printed in only red and blue; by the time the Native American kids came to stay, the lithography had been changed to include a full range of colors, much as Kellogg's had spiced up its former black-and-white boxes with loud yellow and full-color artwork.

Having added a sugar coating to both wheat and rice, Post then decided to try it with their brand of corn flakes. Does that sound vaguely like what Kellogg's was also working on at the time? Post actually had the idea first, but its brand of presweetened Toasties, given the catchy name of Corn-Fetti, did not quite have all the problems worked out. The sugar coating hardened into an inedible shell, and while Post went back to the drawing board to come up with a new formula, Kellogg's got a whiff of what was going on and managed to get Sugar Frosted Flakes (and Katy and Tony) into the market while Post was still mired in the sticky mess of its own devising.

At last Corn-Fetti was straightened out—by making it almost identical to the Kellogg's product—and it was introduced all over again as if it were a new invention, with only the phrase "New! Improved!" to hint that it had already made the rounds once before. In both of its debuts, the advertising for Corn-Fetti was built around a new mascot character who seemed poised to take the spotlight away from Sugar Crisp's three identical bears. Considerably more colorful in appearance and dialogue, he was a patch-eyed, peg-legged pirate known as Captain Jolly.

To date, no evidence has bubbled up to indicate that Captain Jolly ever made much of a splash in animated television commercials. Instead, his ad campaign leaned heavily toward that style of a half decade earlier, comic strips. Magazine ads featured miniature Captain Jolly strips in full color, and especially after the reformulation and reintroduction of Corn-Fetti, the captain and his motley crew starred in their own comic books given away inside the cereal boxes.

Despite his name, Captain Jolly was every bit as much a pirate as his costume suggested. Pirates were something of a fad in the early 1950s, thanks to Walt Disney's first all-live-action feature film, *Treasure Island,* and the upcoming animated classic *Peter Pan*. Captain Jolly, in his printed appearances, seemed undeniably based on Robert Newton's screen portrayal of the lovable villain Long John Silver. In one early magazine ad, Captain Jolly's crew secretly frets because they have accidentally dropped a whole treasure chest of Corn-Fetti overboard. The first mate threatens the ship's cook: "He's mean as a man-eating shark when he doesn't get his corn flakes with the magic sugar coat! Better practice dodging . . . He'll crown you with the sugar bowl if you try to serve him any cereal but Corn-Fetti!" Once the floating chest of cereal is fished out of the drink, Captain Jolly initially punishes his wayward underlings by refusing to let them have one bite of his delicious "treasure," but then his better side comes out: "Ah, dive in! I can't stay mad when I'm eating the crispest cereal on land or sea! Doesn't go mushy, even when you drown it in milk!"

Say, wait a minute—didn't that last slogan of the Captain's sound rather familiar? In the years since, there has been speculation that Corn-Fetti and its shipmates were the direct or indirect inspiration for Quaker Oats' later megahit Cap'n Crunch (which "stayed crunchy, even in milk"). We will be getting to the Cap'n's story later, as it indeed belongs to a different time and place than the Corn-Fetti sea chantey.

Post's Corn-Fetti and its mascot character, Captain Jolly the pirate, have been cited by several historians as the grandfather of Quaker's later hit product, Cap'n Crunch. Captain Jolly was a pirate through and through and often threatened bodily harm to crew members who got into his Corn-Fetti stash.

While a character such as Captain Jolly was a fitting way to sell a sweet treat such as Corn-Fetti, by the mid-1950s Post seemed to be getting a little carried away with the mascot bit. Even Kellogg's and the Leo Burnett agency had the good sense to emphasize characters on products most likely to appeal to kids. No one expected to see a cartoon figure on All-Bran or Special K, for example. Post, however, felt a bit differently, and during 1955–56, was assigning some of the strangest products to the character lineup.

One of the weirdest, although it might be difficult to pick such a front runner, was 40% Bran Flakes. Just the name reeked of its no-nonsense, anti-fantasy theme. Why, then, did Post decorate the box fronts with a pudgy chef with a magic wand, making a bowl of the intestine-cleansing cereal materialize out of a "Magic Oven"? (The oven even had a sign on it designating it as such for anyone who might have wondered.)

Even the chef's connection to 40% Bran Flakes was not as esoteric as the character assigned to push those prewar Post staples, Grape-Nuts and Grape-Nuts Flakes. From your own childhood, you may recall the Aesop's

Fable called "The Fox and the Grapes." So, never mind that Grape-Nuts never had anything to do with grapes. (The name came about because of an initial erroneous belief the human body would turn the nutrients in the cereal into grape sugar.) Post introduced a cunning fox to serve as the Grape-Nuts spokesman, usually depicted running on all fours with a bunch of grapes in his mouth, or conversely, dancing about on two legs while dangling the juicy fruit from one front paw. Either way, the connection between foxes, grapes, and Grape-Nuts was so tenuous that few people paid much attention.

The Grape-Nuts fox did hang around the henhouse long enough to be immortalized in at least one mail-in premium. In 1956, fifty cents would be sufficient to get a set of plastic cereal bowls with the leading Post mascots embossed into them: the Sugar Crisp bears, the Rice Krinkles clown, the Toasties Indian kids . . . and yes, the woefully out-of-his-element fox, too. There was no doubt that the bears were the leaders of the gang, having predated all the rest. Besides the bowls, Dandy, Handy, and Candy were also stamped into a plastic sugar and creamer set and drinking mugs, an honor not afforded to their lesser brethren.

It might seem strange that non-kid-oriented cereals had their own mascots, while Post's new presweetened product for 1957 did not. Alpha-Bits were, as the name implied, shaped like twenty-six different letters, fashioned from oats. To keep them from tasting exactly like every other oat cereal, they were given a "sugar sparkled" frosting, but except for the drooling kid holding out his bowl for them, the box featured no specific logo character. Things would change in the 1960s, but for the time being, the novelty of the cereal shape was the alpha and omega of Alpha-Bits.

Also during 1957, Post began moving its television sponsorship from the old live-action western series to the relatively new medium of TV cartoons. One such series benefiting from Post's input was CBS's *Mighty Mouse Playhouse*, widely acknowledged as the first Saturday-morning cartoon show when it debuted in 1955. The network had bought the Terrytoons studio, which had been churning out theatrical films since the days of silent movies, and packaged the vintage cartoons to reach a new generation. Mighty Mouse was the star, as the title suggested, but the heroic rodent received admirable backup from the smart-aleck magpies Heckle and Jeckle and their nemesis, Dimwit Dog; Sylvester the Fox; Dinky Duck; Gandy Goose; and other ex-movie stars. All the members of this cast pitched in on Post's cereal boxes during this period and appeared in commercials during the TV show. For example, Mighty Mouse could be seen liberating his favorite Post cereals from the clutches of various and sundry villains (usually of the feline persuasion) and leading the audience in singing Post's newest jingle: "Most any cereal is fine with me/As long as you spell it P-O-S-T/'Cause all Post cereals happen to be/A little bit better than any other cereal happens to be."

Although they were not animated for television, another set of cartoon characters were conscripted for Post duty in 1957. Li'l Abner and the other denizens of Dogpatch must have finally grown tired of depending on Cream of Wheat to keep life peaceful, because suddenly their allegiance was switched to Toasties, Grape-Nuts, Raisin Bran, and 40% Bran Flakes. Post used Al Capp's hillbillies in contests, such as one to name the diminutive sweetheart of Li'l Abner's son, Honest Abe (Abner finally having married the determined Daisy Mae in 1952). The box backs were printed with "Amoozin' Li'l Abner Pop-Out Pictures," which could be assembled to give semi-3-D renditions of scenes from the comic strips.

By then, another set of characters had been added to the usually adult-marketed Post Raisin Bran. They were not given names, at least to anyone's knowledge, so they can be described only as "sugar fairies," flitting about the cereal bowl on the box front. With their glowing wands, they resembled chubby relatives of Disney's Tinker Bell, who performed similar stunts to introduce the weekly *Disneyland* TV show. Like the Magic Oven chef and the Grape-Nuts fox, the fairies did not hang around the grocery aisle for very long.

A more lasting part of Post's cast of characters began to emerge during 1957. Some of the account executives at Benton and Bowles came to the conclusion that Sugar Crisp really did not need three bears to move the merchandise, especially when no one could tell them apart in the first place. The temporary solution was to make one of the bears several times larger than the other two on the box front. This was carried over into the animated commercials, where the "large" bear would come to life off the box while his two formerly equal brethren remained immobilized and moribund as background art.

The new character (or, more correctly, the reimagining of an old one) was still not given a name, usually identifying himself as "a Post Sugar Crisp bear." At least his voice was finally crystallized as a young, child-like sound, unlike the earlier years when the three bears' voices changed with each appearance. In another effort to help everyone forget that once upon a time there were three bears, the new single cub condensed the long-running Sugar Crisp slogan to, "For breakfast it's dandy, for snacks it's so handy," completely eliminating the "eat it like candy" part of the rhyme.

The <u>eatingest</u> youngsters just happen to eat Post Sugar Crisp

For breakfast it's dandy—for snacks it's so handy. The darling of the eating set. Plump puffs of muscle-building wheat, sparkled with sugar 'n' honey. Solid with nourishment.

All Post cereals happen to be just a little bit better!

In 1957, Post's ad agency decided that having three bears to advertise Sugar Crisp was unnecessary. As seen on the box in this ad, the temporary solution was to make one bear more prominent and relegate the other two to the distant background.

In the last chapter, we saw how Kellogg's jumped at the chance to sponsor all of the early Hanna-Barbera cartoon series. What that success wiped out from people's memories—which was just fine with Kellogg's—is that Post had actually been the sponsor for the studio's *first* made-for-TV series, *Ruff and Reddy*, in 1957. As would happen so often in the future, the commercials with Ruff (a cat) and Reddy (a dog) featured better animation than the show they interrupted, due to bigger budgets. In some of them, Ruff and Reddy interacted with the Sugar Crisp bear, while in others the intrepid pair had adventures of their own, resolved by judicious distribution of the product.

By 1958, Captain Jolly had been sailing the bounding main for Corn-Fetti for so long that the old salt was beginning to seem a bit waterlogged. Post chose to send the ancient mariner to the old pirates' home and reformulate Corn-Fetti into a "new" product renamed Sugar Coated Corn Flakes. Taking over Captain Jolly's mascot duties was a new character, a humanized ear of corn who introduced himself to his live-action kid co-stars with the following dialogue: "'WOW!' is my middle name! Cornelius W. (for WOW!) Sugarcoat . . . and don't call me 'Corny'! Now, if you'll lend an . . . ahem . . . EAR, you might hear something that'll BOWL you over!" Cornelius's voice was a not-too-close imitation of W. C. Fields, which well suited his deliberately corny jokes.

Another new Post product introduced just as the 1950s were ending was Heart of Oats. Since this cereal was merely a heart-shaped version of Cheerios, just as Alpha-Bits was a sweetened letter-shaped rendition of the same thing, one might expect Post would borrow some of the same selling points General Mills was currently employing. One would be correct, but only partially.

To promote Heart of Oats' alleged strength-giving qualities, a new animated character was introduced: a barrel-chested, muscular beast in a tight T-shirt who gave his name as Linus the Lionhearted. The voice in which he delivered that introduction would have been immediately familiar to the parents of the kids watching the commercials at home. The "tough guy" tones belonged to veteran character actor Sheldon Leonard, who had won renown for his screen portrayals of gangsters in movies and for parodying his persona on such radio programs as Jack Benny's. By the late 1950s, Leonard had become a phenomenally successful television producer, which probably explains how he ended up as a cartoon lion. Several of Leonard's TV series, most importantly *Make Room for Daddy* with Danny Thomas, were sponsored by General Foods and Post, and they would enjoy a cozy relationship throughout the 1960s. When someone at

When Corn-Fetti was reformulated as Sugar Coated Corn Flakes, Captain Jolly was set adrift in favor of a talking ear of corn who introduced himself as Cornelius W. Sugarcoat.

An obscure Post product known as Heart of Oats served as the introduction to the company's legendary mascot Linus the Lionhearted. Linus would have to undergo some major revisions in appearance and personality before rising to his tremendous fame during the ensuing decade.

Benton and Bowles decided Leonard's growl would be a natural fit for a lionhearted lion, he signed up with great good humor.

In Linus's first commercial, he is initially seen demonstrating his strength by using only his tail to sock a would-be lion hunter out of the picture. "I have da strength dat comes from oats," he explains in Leonard's gangsterese accent. "Dat's what 'lionhearted' means. I get my strength from new Post Heart of Oats dat gives ya da nourishment of oatmeal in a delicious new way." At that moment, he spies a young boy wearing a pith helmet creeping through the background scenery. "Dis little kid can show ya what I'm roarin' about," he decides. "Hey, friend!"

The boy is naturally frightened and opens his mouth wide to yell, "HELP! A LION!" This gives Linus the opportunity to pour half a box of Heart of Oats down the screaming kid's gullet, instantly giving the youngster a bodybuilder's physique and the courage to match. "Now you've got da strength of oats from da lionhearted cereal . . . Whaddaya say?" The boy rewards his benefactor with a loud roar, picking up Linus and tossing him away just as Linus did with the earlier hunter. "See wot I mean?" the dazed but satisfied lion says to the audience.

As a cereal falling somewhere between a kids' product and an adult one, Heart of Oats was marketed through these animated Linus commercials as well as live-action spots that featured no cartoon lion (or exaggerated claims of strength) at all. Then there was an interesting combination of the two, in which a live-action housewife is startled to discover the animated Linus in her kitchen. "Are you a people eater?" she asks. Linus assures her that he lives on Heart of Oats and feeds her a spoonful. We aren't quite to the equality phase of such advertising, so the lady does not demonstrate any notable feats of strength, but she does roar at Linus with a facial expression that has to be seen to be believed.

A few pages hence, we will learn what became of Linus after some cosmetic surgery and personality tinkering in the following decade. For now, we must leave him in the newly ferocious housewife's company and examine the third of the three biggest cereal mascot employers. Get ready to salute a general . . . General Mills, that is.

The Big "G" Stands for "Goodness"

General Mills in the 1950s and 1960s

As we saw earlier, while Kellogg's and Post were slugging it out for supremacy, for many years General Mills was considered the pathetic third-place runner. The company had been a late arrival on the scene, and even as the 1950s and the television era began, General Mills was still basically dependent on its two earlier hits, Wheaties and Cheerios. That would have to change if General Mills were going to get down and dirty in the cereal bowl with its rivals.

In 1951, Wheaties developed an association with the Walt Disney Studios that, if not as pervasive as cartoon connections would be a decade later, at least gave kids a reason to gobble down the product. Unlike the later practice, the Disney characters did not appear as logos on the Wheaties box fronts, but they certainly took up all of the available space on the backs. What would otherwise have been throwaway cardboard space became cut-out masks of Disney's biggest stars of the era: Mickey Mouse, Donald Duck, Pinocchio, Dumbo, Brer Rabbit, Bambi, and (from the hit movie of the previous year) Cinderella and the crafty Lucifer the Cat. Simply cutting along the dotted lines and adding string produced semi-3-D likenesses of these animated legends' visages.

For kids who wanted to do more than run around looking like Bambi or Brer Rabbit, Wheaties also offered a long series (running at least twenty-four volumes) of miniature Disney comic books as mail-order premiums. The aforementioned characters, plus additional friends including Goofy, Cinderella's mice friends Gus and Jaq, and Donald's no-nonsense Grandma Duck could be found in adventure after adventure available for fifteen cents per eight comic books. Yes, you Disney collectors out there, read that price and weep.

While the Disney characters sat around the castle and waited for the next call from General Mills, Wheaties went back to promoting sports stars in order to retain its reputation as the "Breakfast of Champions," and Cheerios remained characterless in its traditional bright yellow box. The company introduced Kix, a "round corn flake," but it, too, did not depend on a mascot to help sales. Then, seeing what the competition was up to, in 1953 General Mills leaped into the presweetened cereal market with a mixture of corn and wheat puffs it dubbed Sugar Smiles.

Like the other General Mills products of its time, Sugar Smiles had no mascot character except the generic, grinning, cowboy-suited kid on the front of the box. The product really could have used help, because amid the clamor for Sugar Crisp, Sugar Pops, Sugar Smacks, and Sugar Frosted Flakes, Sugar Smiles was barely able to crawl out of the sugar bowl.

In retrospect, it seems the General Mills ad executives were a bit slow on the uptake, because the company already had one of the biggest stars in children's entertainment in its stable. Cheerios had been sponsoring *The Lone Ranger* on radio since 1941 and on television since 1949, but it was not until 1954 that someone had the brilliant idea of putting the straight shooter's masked face on the boxes. Once such a rather obvious decision was made, Cheerios treated the Ranger in the same fashion Wheaties had approached the Disney crowd: masks of the *Lone Ranger* characters, both heroes and baddies, were printed on the Cheerios and Wheaties box backs, while miniature comic books of the same dimension as the Disney comics were placed inside. The Lone Ranger appeared in person on the box artwork to hawk these premiums, making him the nearest thing to an official mascot that any of the General Mills products had up to that point.

The Ranger's marketing appearances continued to be something of an anomaly among the General Mills lineup, however. In 1954, the product developers took Kix and drenched it in fruit flavors (lemon, orange, and raspberry), calling the resulting treat Trix. Apparently no thought was given to having a character that could be identified with Trix, as the box featured a generic pattern made up of the tri-colored corn balls.

Another product introduced the same year did delve a little more deeply into the copywriters' imaginations. The short-lived Sugar Smiles had grinned its last and was reborn as Sugar Jets. The bright blue and yellow boxes proudly displayed artwork of what could safely be identified as General Mills' first original mascot character, Major Jet.

This is not to say that Major Jet had much more personality than the piece of cardboard on which he was printed. Looking vaguely like a spaceman and even more vaguely like an airplane pilot of some sort, Major Jet was not called upon to do anything more than seem heroic. The boxes also featured two of Major Jet's sidekicks, a pair of cartoon kids who flew about at supersonic speed, apparently powered by nothing but their internal organs hopped up on Sugar Jets. These hyper squirts also appeared in animated form in the television commercials.

General Mills might have kept halfheartedly stumbling around like this indefinitely, but one important development in the company's television sponsorship dragged it kicking and screaming into the world of major character mascots. In October 1955, the ABC network premiered *The Mickey Mouse Club*, an hour-long daily visit into the wonderful world of Disney. General Mills signed on as one of the rotating sponsors, and it immediately became apparent that the company's existing way of pushing cereal just was not going to fit with the shenanigans of Mickey and the Mouseketeers.

It was General Mills' good fortune that the Disney studio had recently established an all-but-top-secret division to produce animated television commercials. Walt himself hated the idea of prostituting the art of animation for such purposes, especially with the cost-cutting measures that had to be taken. His staff of veteran artists were also holding their noses with one hand and their graphite pencils with the other, but the fact was that the studio needed money to help defray the costs of its quality television output. There was also the matter of a little project called Disneyland being built in some former orange groves in Anaheim, California, so the commercial animation division came into being out of necessity. This is where General Mills' first commercial superstar had his origin.

Riding into the picture, and onto the TV screen, came the Cheerios Kid. This character had an antecedent in General Mills' newspaper-comics ads of the late 1940s, in which a baggy-pants boy known as Cheerios

Joe got instant energy from the cereal. However, the Cheerios Kid was destined for a much more impressive career.

The new Cheerios Kid commercials were possibly the earliest, longest-running, and nearest imitation to the Popeye format stretching back to the Wheatena days (not to mention the animated cartoons even before that). The Cheerios Kid was a young boy, originally dressed as a junior cowpoke, who would have to extricate himself from predicaments by a healthy ingestion of Cheerios. When he did so, he would graphically demonstrate his indebtedness to Popeye by flexing a huge bicep with the image of a single Cheerios oat circle glowing inside.

While the general style and format of the commercials would be revised over the years—sometimes with an alarming frequency—the earliest ones generally depicted the Kid as something of a loner. One of the first jingles went so far as to give the Kid's name as Harry, but that was soon discarded in favor of having other characters address him simply as "Kid." The spot that identifies him as Harry put him in a position to observe a nearby locomotive speeding toward a washed-out trestle, thereby endangering the occupants. Perhaps Harry the Kid could have thought out his response a bit further if he had had the time, but as it stands, he swallows his Cheerios and then stops the train with a mighty punch that destroys the engine and flips the aged engineer out of the cab and onto his head. Just how wrecking the train is an improvement over letting it plunge down a gorge is one of those cartoon conventions that must not be questioned.

Since the commercials were being animated by Disney for airing during a Disney TV show, it seemed only logical to team the Cheerios Kid with the irascible Donald Duck for another series of spots. In one, Donald and his nephews visit the beach, where the obstinate quacker resolutely ignores the signs warning swimmers about the shark-infested waters. Donald soon finds himself surrounded by the toothy predators, while the nephews spot the Cheerios Kid working as a lifeguard nearby.

As the Kid pours a box of Cheerios down his throat, the narrator cheerfully assures the viewers, "Sure, the Cheerios Kid can save Donald, and here's why! A Cheerios breakfast gives you protein to help build and maintain strong bodies! Real power protein, plus Vitamin B1 to give you GO POWER!" As the off-screen singers strike up the jingle, "Yes, he's got GO power/There he GOES!/He's feeling his Cheerios!" the Kid rockets to the rescue and slams the shark in the jaws.

A similar plot unfolds in another commercial in which Donald and his nephews are hiking in the woods and the fowl runs afoul of a bear.

The Cheerios Kid was introduced during commercials on *The Mickey Mouse Club* in 1955. For one series of adventures, he had the unenviable task of acting as protector to that most foul-tempered of waterfowl, Donald Duck.

This time, the Cheerios Kid is a ranger stationed in a fire tower, and with a mouthful of Cheerios, he dispatches the bear to the happy hunting grounds. Each spot ends with the Kid flexing his oat-enhanced muscle as he sings, "If you know your oats you too will go/For the power breakfast, Cheerios!" For whatever reason, the Kid was probably the most major cereal advertising character to not be featured on the front of his own product's boxes; as a sort of consolation prize, comic-strip renditions of some of his adventures with Donald were printed on the backs.

General Mills' other main product, Wheaties, also jumped on the Disney bandwagon during the *Mickey Mouse Club* era. Instead of hiring the animated commercial division, though, Wheaties took the unusual route of bringing in master puppeteer Bil Baird. Although his name is less familiar today, in the mid-1950s Baird occupied a position roughly equivalent to Jim Henson's later celebrity with the Muppets. Baird and his wife, Cora, primarily worked in the medium of marionettes, with occasional hand puppets thrown into the mix. The talents of Baird and his staff were evident in the amazing versatility and range of his stringed thespians.

Since Kellogg's was roaring around the cereal-aisle jungle with Tony the Tiger, Baird supplied Wheaties with a character who would have occupied

Kellogg's might have had Tony the Tiger, but General Mills had Champy the Lion to shill for Wheaties during commercials on *The Mickey Mouse Club*. Champy and all of his furry friends were creations of master puppeteer Bil Baird.

an even higher position in the beast hierarchy: Champy the Lion. The Champy commercials took the form of miniature dramas playing out something like a cross between a Warner Bros. *Road Runner* cartoon and Disney's *Song of the South* treatment of the Uncle Remus tales. Most of them opened with a chorus singing the jingle: "Wheaties time, Wheaties time/Breakfast of champions time!" Champy would then chime in with his deep voice: "The way it works is magic/The things it does for you/ Helps you be a champion of/Whatever you want to do!"

In one adventure, Champy is out on the lone prairie, sharing a campfire with his donkey-stagecoach-driver friend. On a cliff high above, the villainous Mr. Fox and his buzzard accomplice are preparing to drop a huge boulder on Champy's crowned head. "Once we get his Wheaties, we'll be big, strong champions just like him," the buzzard croaks. Mr. Fox instructs the vulture that when he reaches a certain spot at the bottom of the cliff, he will whistle as the signal to drop the boulder. In true cartoon fashion, friendly Champy unexpectedly walks over to say hello, causing the fox to run into the boulder's path and get flattened like an all-Wheatie pancake.

Another plot is cooked up by a mouse dressed like a prizefight manager and his partner in crime, a bear with an expression even more stupid than one might think possible. The mouse tries to explain to his addled assistant that the Wheaties box he is holding is a decoy and contains a bomb. (The fake package is a hilarious parody of the real box, with "WEETYS" scrawled across the front and a stick-figure drawing approximating the baseball player found on the genuine article.) The mouse's plan is for the bear to distract Champy's attention long enough to switch boxes.

Somehow, the bear manages to do just that, but while he gets lost in admiring Champy's many sports trophies on the shelf, the lion secretly switches the two boxes' positions back again. On cue, the mouse yells from outside, "Call for Mr. Bear!" causing the dope to grab the decoy package and run off with it, triggering a huge off-screen explosion. We never do see the remnants of the two would-be villains.

Like the Cheerios Kid and virtually every other General Mills commercial character until the early 1960s, Champy and his friends did not appear on the Wheaties boxes. The company did market an array of premiums, however, including a pair of well-made Champy and Mr. Fox hand puppets and accompanying cardboard theater. Bil Baird stepped out from behind the scenes to push these items in a commercial of his own, giving him a chance to remind the marionette Champy, "I designed these puppets, just like I designed YOU!" (Even the healthful qualities of Wheaties could not prevent Champy from eventually meeting his maker.)

Encouraged by the success of the *Mickey Mouse Club* commercials, in 1958 General Mills introduced three new characters—two of whom represented cereals that had never before had their own mascots. The one exception was Sugar Jets, which jettisoned the flying kids and replaced them with Mr. Moonbird.

Mr. Moonbird was one of the more unusual mascots to ever march across a kitchen table. He was presented as a beatnik jazz musician, leading one to suspect that he doused his Jets with espresso rather than milk. With an antenna growing out of the top of his head and dialogue leaning heavily on, "Crazy, man," he would have looked right at home on the Sunset Strip. Scott Bruce has accurately commented, "Unlike Major Jet, who held that space could only be reached with the right hardware, Moonbird suggested the final frontier was as much a state of mind."

Kix, which had been getting along fine without a character, introduced Eager Beaver. Like Mr. Moonbird, the beaver was not so much a box-front logo as an animated commercial spokesman. The third cereal to receive a character that year was Trix, which introduced a rabbit. Uh, no—not *that* rabbit. He comes in later. This was Magic Rabbit, an escapee from the vaudeville stage who was made up of equal parts of the Mad Hatter and the March Hare. He did not last long, but at least he managed to get his rabbit's foot in the door—which would soon prove to be very lucky for Trix.

A couple of brand-new cereals were also introduced in the late 1950s. At least they were promoted as new, ignoring the fact that their creation mainly involved some tinkering with existing products. Having already added fruit flavors to Kix to birth Trix, the same corn puffs were now coated in chocolate to be unveiled as Cocoa Puffs. The name did not seem to spark any particular image in the ad executives' minds, so the best characters they could come up with were three children who enjoyed playing "train" and chanting out the cereal's slogan: "Puff, puff, Cocoa Puffs/ Chocolate-flavored Cocoa Puffs/Cocoa Puffs for energy, Cocoa Puffs for fun/Chocolate-flavored cereal, yum yum yum!" That was about the extent of their characterization, but it would have to suffice for the next four years until something a bit more cuckoo came along.

Blandness was hardly a problem when General Mills gave Cheerios a thick sugar frosting and introduced the result as Frosty O's. This time, the ad agency had an immediate tie-in with the cereal's name, and introduced the public to Frosty the Polar Bear. One of his first commercials tied the two related products together, as Frosty and the Cheerios Kid teamed up as (respectively) barker and strongman of the "Circle O Sideshow." The Kid demonstrates his prowess on the ring-the-bell game, knocking the bell clear out of sight. Frosty, pedaling about the circus grounds on his unicycle, describes his new cereal as "super circles, brand-new, cookie-crisp sugar-charged oats." With the debatable claim, "They're sugar-charged, so you'll be SUPER-charged!" he demonstrates their potency by turning some juvenile spectators into a skilled team of acrobats. Even after this introductory spot, while the Kid went back to his usual adventures, Frosty continued to hang around the circus to gin things up with his sugar-laced pep-up potion.

All of this was leading up to the 1959 debut of General Mills's first true superstar. (Okay, so the Cheerios Kid had super strength, but he never attained the level of true front-of-the-box stardom.) The Trix commercials had been trickling along for five years with assorted generic children extolling the fruity puffs' flavor, when suddenly a spot appeared that took a completely different approach. In fact, from its opening lines, viewers would have had no immediate clue as to just what it was about.

Frosty the Polar Bear was on hand to introduce the new sugar-coated version of Cheerios known as Frosty O's in 1959.

An animated white rabbit standing in front of a heap of fresh produce began talking directly to the audience:

> RABBIT: I have a problem. I'm a rabbit, and rabbits are supposed to love carrots. But . . . yecchh . . . what I love is Trix! Trix is the corn cereal with fruit flavors, and they come in fruit colors: raspberry red, lemon yellow, and orange orange, all mixed together in one big box. They're crispy, crunchy and mmmmmm, boy, do I love 'em! I could eat 'em for breakfast, and lunch and supper, and even after school . . . if I went to school.
>
> (A little boy approaches and grabs the box out of the rabbit's hand)
>
> BOY: Silly rabbit . . . Trix are for KIDS!
>
> RABBIT: (to audience) That's my problem. Trix . . . are for kids.
>
> (The boy has now joined a little girl at a breakfast table)
>
> GIRL: When I grow up, I'm going to have a whole houseful of Trix.
>
> BOY: Yeah!
>
> GIRL: Don't you just love Trix's fruit colors?

BOY: Yeah!
GIRL: And that fruit flavor!
BOY: I like them better than anything.
(The girl notices the rabbit standing next to the table with a mournful expression)
GIRL: What's the matter with HIM?
RABBIT: I'm sad because Trix are for kids.

And so began an ad campaign that is still running more than fifty years later. What was it that made audiences take the rabbit to their hearts more so than his predecessors? Much of it can probably be attributed to his voice, supplied by character actor Mort Marshall. The rabbit did not sound like any other character in cereal advertising, with his tones alternately whiny, pleading, and (when faced with a box of Trix) wildly ecstatic. The rabbit could possibly also be considered the first cereal mascot who personified the "antihero," a figure that was rapidly gaining popularity through the movie characterizations of Marlon Brando and James Dean. The rabbit never won his contests, and had he managed to actually get a taste of Trix, it would not have given him super strength or any other preternatural abilities; it was purely a sensual experience for

The Trix Rabbit looked a bit different in his first commercial in 1959, but his problem was the same as always: he craved Trix, but "Trix are for kids." More than fifty years later, the long-eared rabbit is still facing the same dilemma.

him. Perhaps, out of all the real and animated people, animals, and more fantastic creatures, this rabbit was the most human of all.

Inasmuch as General Mills was now developing a cast of characters that, if not yet exactly in the same league with the Kellogg's crew, at least had the competitive potential, the company started looking jealously at its rival's TV success. The linkage between Kellogg's and the Hanna-Barbera output was growing stronger every day, and the folks at General Mills wondered how they could be a part of the same market. The company needed an animated show to sponsor to shoot it into the TV toon-time stratosphere—and what better star to accomplish that goal than a flying squirrel?

Through a series of negotiations and complicated contracts, General Mills aligned itself with relatively unknown cartoon producer Jay Ward. (His only previous success had been *Crusader Rabbit* ten years earlier.) Ward and writer Bill Scott had been trying to sell a series they called *Rocky and His Friends*, built around "that jet-age aerial ace" Rocky the Flying Squirrel and his dense but loyal friend Bullwinkle the Moose. The details of the General Mills/Jay Ward agreement are thoroughly dissected in Keith Scott's definitive history of the studio, *The Moose That Roared*, but as far as our discussion is concerned, they broke down into two primary caveats: the animation for *Rocky and His Friends* would be physically executed for a bargain-basement price in Mexico, and General Mills would own the television rights to the show in perpetuity.

Well, maybe the first of those two rules had the most flexibility. Over Ward's objections, General Mills was so cost-conscious they would have had the show drawn on three-ring notebook paper and colored with crayons if it were possible. However, when it came to the commercials featuring the cast plugging away for Cheerios, Jets, Cocoa Puffs, and Trix, General Mills wanted them to be flawless, so those spots were to be animated by Ward's own staff in Hollywood. As we saw earlier, there was a difference in style between the commercials produced by Hanna-Barbera and the shows they interrupted—but at least both elements were coming out of the same studio. In the case of *Rocky and His Friends*, the commercials were so slick, and the rest of the show so crude, the program seemed to be suffering from multiple personality disorder.

While Rocky, Bullwinkle, and their friends (and sometimes enemies) appeared in commercial after commercial for the General Mills products, the company stubbornly stuck with its refusal to clutter up its well-designed packaging by featuring them as mascots. One minor exception was when genius dog Mr. Peabody and his pet boy, Sherman, appeared briefly on the boxes of Wheat Hearts, an unsuccessful attempt to copy Post's Heart of Oats.

Rocky and His Friends began airing weekday afternoons on ABC-TV in November 1959. Even with its visual shortcomings, the scripts and voice acting were so hilarious that it quickly caught on with both children and their parents. General Mills certainly had reason to laugh about it, and soon decided they wanted another show to maximize their exposure. With Ward's entire resources and then some tied up in meeting the impossible deadlines for *Rocky*, General Mills turned to another newborn studio, Total TeleVision Productions, for its next series, which was executed under the same conditions as Ward's series: the animation would be produced in Mexico (except the commercials), and General Mills would forever and after own the TV rights. The new show was humorously titled *King Leonardo and His Short Subjects*.

This series premiered in October 1960, and served as a vehicle for introducing General Mills' latest cereal, Twinkles. Since there was nothing particularly innovative about the product itself, the packaging came with a gimmick. By tearing along a perforated line, the back of the box would open up into a three-page storybook. The eponymous star of the stories, and the commercials promoting them, was Twinkles the Elephant.

Twinkles was featured in brief stories that were incorporated into the *King Leonardo* show, and their gentle, preschool style was an ill fit for the rest of the program's sometimes slam-bang humor. The commercials followed the same pattern, although with occasional lapses into violence that seemed out of character for the mild-mannered pachyderm. For example, let's take the dialogue accompanying one of the first Twinkles commercials:

> TWINKLES: (buzzing into view by using his tail as a helicopter) Hi, mind if I drop in? I'm Twinkles, the magic elephant! I tell a story on every box of new Twinkles, the only cereal in the storybook package. Tear the magic line, and your storybook package opens! I've got so many stories to tell . . . Why, once I remember how I got trapped in a deep pit. I had to rev up my rotor-tail to escape! (He demonstrates his tail's unique abilities.) Another time, I was chased by elephant-eating sharks! Then, I had to turn my magic trunk into a harpoon gun. (Twinkles blasts the off-screen sharks into the afterlife.) Oh, you'll like my Twinkles stories . . . and wait 'til you taste Twinkles! Delicious stars with the goodness of energy oats and corn! Only Twinkles has the storybook package; tear the magic line and there

I am with all my friends! Look for us on Twinkles, the only cereal in the storybook package.

In the early ads, as well as in the mini-stories similar to the ones on the boxes, narration and all voices were performed by George S. Irving, whose distinctive tones would become a trademark of Total TeleVision's output. (Some readers might also know him as the Heat Miser in the classic animated special *The Year Without a Santa Claus*.) Although several sources have referred to Twinkles as a pink elephant, casual examination

A pair of Twinkles the Elephant's gentle adventures, with all voices performed by George S. Irving, were released on a Little Golden Record in 1961. Twinkles was one of the first cereal mascots to be marketed in retail stores in a major way.

of the many pieces of merchandise produced shows very clearly that his color was always orange.

It was somewhat unusual for a cereal mascot to be the subject of toys sold at the retail level, in addition to mail-away or free-inside-the-box premiums, but Twinkles made that big jump from cereal shelves to toy shelves, at least for a couple of years. There were Twinkles jigsaw puzzles, storybooks, and Little Golden Records, although most of the items gave absolutely no hint of their grocery-store connection. About the nearest they got was a board game called "Twinkles and His Trip to the Star Factory," which subtly referenced the shape of Twinkles (the cereal, not the elephant).

Sugar Jets were still hanging around in midair, although their sales weren't breaking any sound barriers. Around 1960 the name was changed simply to Jets, and the box-front character became known as Johnny Jet. He was a young boy in an aviator's uniform, and his rudimentary facial features strongly prefigured the ubiquitous smiley face of a decade later. Johnny Jet appeared in commercials explaining how consumption of the cereal helped him win every contest, but the same theme was carried out more memorably when Rocky and Bullwinkle were brought into the campaign to fill his spot in the spots.

In one vignette, Bullwinkle is competing in a bicycle race and losing fast. Rocky arrives, proclaiming, "Never fear, Jets are here! The cereal with jet energy . . . helps ya win!" Bullwinkle wolfs down a bowl—or mooses down a bowl?—and passes the other competitors with such supersonic speed that he comes in first, second, *and* third in the race. In another adventure, Bullwinkle is a prizefighter who is being pummeled into a booby prize by the brutish champ. Again arriving with Jets, Rocky feeds them to his partner, who spins his arm like an airplane propeller and disintegrates the palooka in a clap of thunder. "Win with Jets!" Bullwinkle advises his viewers.

In 1961, the Trix Rabbit (he still had no other name assigned to him and never would) was afforded the honor of finally appearing on the front of the box. Around the same time, his commercials began increasing the magnitude of his legendary schemes to snag a taste of the cereal by any means necessary. In one rather drastic move, he enlists in the U.S. space program and volunteers to be shot to the moon, where he assumes he will be able to gobble Trix with no outside interference. Once his space capsule crash-lands, he discovers his glass helmet prevents him from being able to get the spoon into his mouth. If that were not bad enough, his box is snatched away by two moon kids who have wheels instead of legs, and who squeak, "Silly rabbit, Trix are for kids" in voices sounding like Alvin and the Chipmunks.

In another plot, the Rabbit declares, "If I can't get Trix any other way, I'll make it myself!" He fashions a cereal-manufacturing machine, but the only box of Trix it produces is quickly grabbed away by a nearby boy. In frustration, the Rabbit kicks his own machine, which then spits out a card with the "Silly rabbit" line. In a still later spot, the Rabbit enlists the help of a "marvelous cereal hound" to sniff his way to Trix. (The dog is played by George S. Irving and says "bow wow" in English instead of barking.) The Rabbit is infuriated when his hound instead leads him to a pile of wooden sticks. "STICKS?" he explodes, throwing a tantrum. "I wanted TRIX! TRIX!" We could hear it coming: the dog scolds him, "Silly rabbit, Trix are for KIDS!"

Have you noticed General Mills seemed to be missing one important cereal component in its lineup? Whatever happened to good old corn flakes, in order to compete with Kellogg's and Post? In 1963, General Mills shucked its veil of secrecy and introduced Country Corn Flakes, the big selling point of which was, "They don't wilt when you pour on milk." The choice of a live-action mascot character was so logical that one wonders how the competition managed to miss the opportunity: a scarecrow that looked uncannily like Ray Bolger's portrayal in *The Wizard of Oz*. As a true expert on corn, the Country Corn Flakes Scarecrow could speak with authority, but after a few commercials and a handful of appearances on the box, he was sent back to the cornfield. Country Corn Flakes' most well-remembered commercial was an animated mélange featuring no continuing character. It bordered on the surreal, with the farm couple from Grant Wood's *American Gothic* painting singing "New Country Corn Flakes, New Country Corn Flakes" over and over with deadpan expressions (at one point switching a single line to "Please buy our corn flakes"). As this deliberately monotonous sound track droned on, animated farm animals participated in simply indescribable antics, ending with a pig hatching a chicken that pops out of the egg and moos. Was "surreal" the right word for it?

Things were getting a bit loonier in the Cocoa Puffs department too. The three kids who had been serving as spokespersons since the cereal's introduction were now as old hat as the old hat the eldest boy wore, so without warning, a new campaign was suddenly sprung on viewers. In the first entry of this new series, an unseen announcer addresses the audience: "Young friends, this is a cuckoo bird." Said bird does not look very cuckoo, except for his outsized beak and striped coat, and he sedately strolls along humming a remnant of the soon-to-be-forgotten "Puff, puff, Cocoa Puffs" jingle. The announcer then sets out to demonstrate how to set the bird off. A table with a bowl of Cocoa Puffs is placed directly in the bird's path, and with a volley of mixed sound effects including an alarm clock, one of the most enduring ad slogans of history makes its debut: "WOW! I'M CUCKOO FOR COCOA PUFFS! CUCKOO FOR COCOA PUFFS!" the bird screams maniacally. General Mills knew it had something special with this birdbrained bird, because his picture on the front of the box was introduced at the same time as his first commercial—a truly rare event for the company.

The voice of the cuckoo, although distorted beyond all recognition, belonged to beloved New York children's TV show host Chuck McCann. For the first few commercials, McCann performed alone, and the bird had no name. With the routine established, it was time to introduce some supporting characters, so into the picture came an elderly cuckoo bird known as Gramps. His voice was an imitation of veteran movie and radio actor Lionel Barrymore, supplied by Allen Swift, who was New York's answer to the West Coast's voice actor Mel Blanc. (Coincidentally, Swift was also a local kids' TV host in the Big Apple, but that role was definitely secondary to his long career in animation.) Since Gramps constantly referred to his young offspring as "Sonny," it became the cuckoo's name by default and remains so today.

The Gramps/Sonny commercials displayed a streak of cruelty not normally considered a viable method of selling cereal to children. Gramps was all too aware of his grandson's psychological shortcomings, and would twist that knowledge for his own advantage. For example, we might see the two of them on a boat in the middle of the ocean, stranded by a broken propeller. While Gramps remains topside, Sonny dives in to try to repair the faulty apparatus but with no success. Chuckling evilly, Gramps mutters he knows how to fix things, and he pours Cocoa Puffs down Sonny's air hose and fills up his glass diving helmet with the sweet treats. This has the expected effect of causing Sonny to go berserk, and his gyrations form a whirlpool, engulfing the entire ocean. After he calms down, he shakily asks Gramps if the propeller is fixed, setting up the punch line: the former ocean is now a dry expanse of desert, and Gramps chortles, "We don't need it—now we can WALK home!"

General Mills was getting into the Saturday-morning television realm in a big way by this time. *King Leonardo and His Short Subjects* had lasted from 1960 to 1963 and was quickly followed by two more Total TeleVision series, *Tennessee Tuxedo* (1963) and *Underdog* (1964). (In *Underdog*, Allen

Sonny the Cuckoo became the mascot for Cocoa Puffs in 1963. Just the sight, or even the mere suggestion, of a bowl of the cereal would cause him to lose control and go on a rampage, leaving destruction in his wake.

This early comic-book ad for Lucky Charms reads like the storyboard for one of the first commercials. Lucky the Leprechaun was far from the benevolent elf he later became; in his early appearances, he truly became angry when kids managed to get the cereal away from him.

Swift got double mileage out of his Gramps/Barrymore impersonation by using it as mad scientist Simon Bar Sinister.) In September 1961, *Rocky and His Friends* had moved from the ABC daytime schedule to a prime-time slot on NBC, where it became known as *The Bullwinkle Show*. General Mills had followed, and stayed with the moose and squirrel as they returned to Saturday mornings for a final season in 1963–64. Once that series ended, Jay Ward immediately jumped into a new Saturday-morning show, *Hoppity Hooper*.

The fact that General Mills ultimately owned the broadcast rights to all of these shows meant after their initial network runs were over, their various openings, closings, and components could be split up and reedited for syndication in dozens of different ways. This led to much confusion among young viewers, who naturally assumed that, say, *Underdog* and *Rocky and His Friends* were produced by the same cartoon studio because eventually their parts were all seen in the same half hours. General Mills did not help matters any by mixing and matching the characters indiscriminately in its premium offers. One mid-1960s item was a plastic "coloring cloth," which could be reused time and time again by rubbing off the crayon marks. The center of this sheet pictured Underdog and the Trix Rabbit, while around the borders were all the other Jay Ward and Total TeleVision casts.

When it was time to introduce a new cereal in 1964, General Mills came up with yet another hit character who would prove to have the same lasting power as the rabbit and Sonny the cuckoo bird. Lucky Charms' big selling point was its flavored marshmallow bits "mixed right in with the cereal," as the ads described it. The new product's animated representative was Lucky the Leprechaun, and while the plots of his commercials stayed the same for the next forty years, their execution certainly changed.

From Lucky's very first appearance, he was seen eluding the clutches of various and sundry children. " 'Tis luck to catch a leprechaun, but o' course no one can!" he cackled in his traditional Irish brogue. In later adventures, the hide-and-seek game seemed to be a much friendlier rivalry between Lucky and the kids, but in the early commercials, Lucky was a considerably more callous character, often endangering the kids' very lives in his attempts to get away from them—leading them into caves, onto precipices, and similar perils. The tables were always turned, of course, and the kids would manage to grab Lucky and his box of Lucky Charms, while the sprite struggled and glowered as if he would like to use his blackest magic to fix their little red wagons.

By the mid-1960s, Lucky the Leprechaun had been redesigned and his personality revamped into a much friendlier character. This design of the Lucky Charms box would last into the early 1980s.

Eventually, just as the Cocoa Puffs commercials were softened by getting rid of Gramps after viewers complained that he was too cruel, the Lucky Charms plots were revamped, and Lucky was redesigned to be a much more winsome elf, with a happy voice supplied by former radio child actor Arthur Anderson. He would still use his magic to attempt a quick getaway after every attempt to capture him but usually failed at the last moment through some sort of Wile E. Coyote–type malfunction. This tended to make him more sympathetic, and he continues his magical antics to this very day.

With most of General Mills' advertising now airing on Saturday mornings, and with the ad agency's newfound appreciation of wild humor, the Cheerios Kid commercials were revamped into a wackier style than their previous "this is to be taken seriously" format. By now the Kid had

By the 1960s, the Cheerios Kid had evolved a long way from his primitive roots in 1955. Some things never changed, though: eating Cheerios still consistently gave him a bulging bicep with a single oat ring glowing in the center.

acquired a girlfriend, the ultracute Sue, who steadfastly hung with him no matter what lamebrain situation he got them into. Now the mini-adventures resembled their *Popeye* origin even more than before, as the Kid would be unable to extricate himself from his latest predicament until Sue channeled Olive Oyl and came to the rescue with a bowl of Cheerios. (Once again, it's time to wonder why Sue didn't just eat the stuff and handle the problem herself, but we'll be getting back to that question several dozen pages from now.)

Another element added to the Kid/Sue combination was the punch line, where she would vainly try to remind the Kid to give her an affectionate kiss as payment for her help. Like all true heroes, though, the Kid's mind was elsewhere.

SUE: (puckered up) Aren't you forgetting something, Kid?
KID: (studying the box) Oh yeah . . . the Big "G" stands for "Goodness"!
SUE: (giving him a mighty slap) Also "girls"!

Hmm, maybe the reason Sue didn't eat Cheerios is because she didn't *need* their artificially endowed strength. Anyway, as the humorous style of commercials continued, the Kid and Sue became almost deadpan in their delivery, leading one to suspect the producers had the *Peanuts* comic-strip characters in mind when writing the scripts and directing the anonymous child actors who supplied the voices. (In fact, in at least one or two commercials, the Kid's voice sounds suspiciously like Peter Robbins, the original Charlie Brown voice.) One of the best scripts from this period had the pair on safari in the jungle:

KID: Look, Sue, a jungle swamp.
SUE: Shall we cross it, Cheerios Kid?
KID: What could possibly stop us?
(A giant alligator suddenly rises from the muck and confronts them)
GATOR: Hmm, how about a nasty old alligator?
SUE: He has a point.
KID: Down, Sue! I'll handle this!
SUE: But you forgot your . . . (the Kid's battered form lands at her feet) . . . Cheerios. Big G, little O means GO with Cheerios! Tasty O's of oats toasted crisp all through; a Cheerios breakfast is packed with muscle-making protein!
KID: (flexing his oat-illustrated bicep) And go, go, GO! See ya later, alligator!

(He bashes all the alligator's teeth out, leaving him with a shriveled snout)

KID: Remember, the Big G stands for goodness! Big G, little O means GO with the goodness of Cheerios!

SUE: (puckering) Didn't you forget something, Kid?

KID: Right. . more Cheerios!

SUE: (dejectedly) I'll never smile again.

TOOTHLESS GATOR: Thath's my line.

For another series of commercials, the Cheerios Kid was separated from Sue and teamed with the recently unemployed Bullwinkle the Moose, whose own show was now in reruns. Most of these featured the Kid trying to train the antlered klutz in some sport or another but not succeeding until Bullwinkle ate Cheerios and developed a Cheerio-displaying bicep of his own. These would alternate with other spots in which Bullwinkle performed alone, but in all of them the message was the same: Bullwinkle would gain strength and stamina from Cheerios, but would remain as clumsy as ever, so his newfound strength only enabled him to crash into the scenery harder. He constantly reminded viewers, "GO with Cheerios . . . (CRASH) . . . but watch WHERE yer goin'!" These commercials ran for the rest of the 1960s and were also adapted into a long series of ads in comic books. Oddly, in the comic strip versions, the Cheerios Kid was replaced by Rocky, who was otherwise absent from the TV scripts.

After producing one enduring hit character after another in the first part of the decade, by 1965 General Mills seemed to be badly in need of some of Lucky the Leprechaun's magic, because there was a string of mascots that could hardly even be considered second string—in fact, they were out-and-out losers. Twinkles the Elephant packed his trunk and retired to wherever ex-mascots go, but the cereal bearing his name was still on the market. His extremely weak replacement was the Twinkles Sprinkler, a daffy fireman whose only shtick was to "sprinkle" the cereal with frosting from his fire hose . . . not exactly the most logical product-character tie-in. Sugar Jets—yes, the "sugar" part of the name was reinstated—jettisoned Johnny Jet and hired Goggol, an alien being whose hyperactivity indicated that on his planet, the inhabitants might have discovered an intravenous use for the product.

Wackies was another cereal mixed with marshmallow bits, but unlike Lucky Charms, these bits were banana flavored. This was the ad agency's cue to introduce a young boy and his pet gorilla, neither of whom was given a name, to play up the wackiness of Wackies. (The gorilla's gravelly

In TV commercials of the mid-1960s, Bullwinkle was teamed with the Cheerios Kid for a series of sporting contests, but in the comic-book adaptations of these scripts, Rocky the Flying Squirrel took the Kid's place.

voice, according to some historians, was supplied by longtime character actor Lionel Stander.) Like Alice after her trip to Wonderland, audiences seemed unwilling to accept pure, unadulterated nonsense, and the boy, gorilla, and cereal were mercifully short-lived. Corn Bursts was another forgotten cereal of the period, promoted in commercials by Hattie the Alligator and a pith-helmet-wearing hunter whom Scott Bruce referred to as Colonel Cornburst. And it gets worse: it might be possible to think of a more anti-health-conscious name for a cereal than Sugaroos, but it would be difficult. The obscure Sugaroos mascots were the outer-space Floops, who obviously came from the same neighborhood in the solar system as Goggol, since a taste of Sugaroos made them "floppity flippity floppity Floops." Are you beginning to see how creating a memorable cereal mascot is not as simple as it might seem?

It could be just as tricky to tinker with an established cereal brand. Frosty the Polar Bear was still giving Frosty O's a blizzard of promotion in 1963–64, but his voice had radically changed from its original deep tones to an impersonation of 1930s–1940s comedian Frank Morgan, performed by Larry Storch. (Like Allen Swift's Lionel Barrymore, Storch recycled his Morgan imitation by using it as Mr. Whoopee in the *Tennessee Tuxedo* cartoons.) By 1965, Frosty had gone back to his polar ice cap and was replaced as Frosty O's spokes-beast by Tennessee Tuxedo's good-natured, dim-witted pal Chumley the Walrus.

Why, you may ask, would General Mills choose the sidekick instead of the main character for a mascot? There are no certain answers, but one educated guess would be the ad agency did not want to pay Don Adams's fee for performing Tennessee's voice in the commercials, as he did for the cartoon series. Instead, when the entrepreneur penguin did appear in the Frosty O's commercials, Larry Storch imitated Adams's delivery as closely as possible. Chumley was the box-front logo character—but only for about a year.

After Chumley, Frosty O's were advertised by a nameless burglar character whose only motivation was that he craved doughnuts and that Frosty O's "tastes like little sugar-frosted doughnuts." A pack of Keystone Kops caricatures were responsible for bringing him to justice in each commercial, but in the long run, crime did not pay—for either the burglar or Frosty O's. After a mercifully short year, the burglar was replaced by a trio of superheroes known as the Energy 3.

By this time, the wacky Saturday-morning cartoons of the early 1960s had been supplanted by a new breed, which children's TV critics derisively labeled the "weirdo superheroes." Trying to fit in with this genre of programming is no doubt what inspired the Energy 3, following as they did in the heroic footsteps of the Fantastic 4 and the Super 6. (If they had all gotten together, they could have called themselves the Unlucky 13.) Actually, the Energy 3 was probably the least heroic of any of them: composed of Oat Man, Sugar Man, and Milk Man, only the first of the three looked anything like a traditional muscle-bound superhero. The other two sported capes and could fly, but Sugar Man looked like a top-hatted Jack Frost and Milk Man looked like—well, a milkman. Combining their powers caused Frosty O's to give its consumers super strength and energy, although not to the Popeye degree of the Cheerios Kid.

Speaking of the Kid, his adventures, too, were influenced by the "weirdo superhero" genre, which had also spawned the "weirdo supervillain." The Cheerios commercials of the late 1960s dropped their earlier self-mocking style and took themselves seriously, even as their situations grew more outlandish (and repetitious). In each one, Sue would be happily minding her own business when she would suddenly be attacked by a freakish creature of another world: "Help! The Rope Man's got me!" "Help! It's the Sponge Man!" The Kid, seeing what dirty work was afoot, would gulp down his Cheerios and then dispose of the bad guy permanently. Since the Rope Man was literally a humanized coil of string, the Kid's Cheerios-powered garden shears went to work and left the baddie as a pile of clippings on the sidewalk, with Sue jumping rope with what was left. A similar fate awaited the Sponge Man, whom the Kid tore into tiny pieces, making them even tinier by squeezing the remaining water out of them. Yes, if you wanted to be able to kill your enemies with impunity, Cheerios was the breakfast for you.

It may be no coincidence that this new attitude toward adventure and villains came about during the escalation of the United States' involvement in Vietnam. While documenting the Jay Ward Studio's history, Keith Scott found multiple examples to back up the claim that General Mills' hierarchy was the most conservative and superpatriotic in the corporate world (Ward had constant battles with them when it came to mocking American history in the Peabody and Sherman cartoons, to name one). It is conceivable that the Cheerios Kid using sheer force to annihilate his adversaries represented General Mills' "nuke 'em" philosophy toward international diplomacy.

By contrast, the mild-mannered Trix Rabbit's never-ending schemes to snare a bowl of the fruit-flavored manna continued even among all his newer brethren in the cereal aisle. By the late 1960s, he was adopting

General Mills offered this set of magnets depicting some of the Trix Rabbit's more creative disguises. By means of this subterfuge, he would occasionally manage to get a spoonful of Trix into his mouth before accidentally revealing his identity.

General Mills' Kaboom contained more than twice the vitamins of other cereals, so to prove the point, the mascot clown would feed it to his fellow circus performers, changing them from klutzes into champions.

various disguises in order to filch a taste. In a rather strange concept, in some of these commercials it was his actual image on the front of the box that would come to life and grow to adult size to aid in his deception. He appeared as a milkman, a balloon seller in the city park, a litter collector, and dozens of more hastily improvised costumes. It could reasonably be argued that his ruse did pay off, because more often than not he would actually manage to get a single spoonful of his favorite treat into his mouth. His resulting ecstasy would cause his long ears to pop out, revealing his identity:

BOY: It's the Rabbit! And he ate some Trix!
GIRL: Silly rabbit, Trix are for kids!
RABBIT: (winking at the audience) And sometimes for tricky rabbits!

General Mills ended the 1960s not with a whimper, but a kaboom. That was the name given to the last new cereal to hit the market for the decade: Kaboom, and for some reason, the logo character was a colorful circus clown. (The cereal pieces were multicolored and shaped like grinning faces, making the connection at least slightly logical.) Kaboom was deliberately formulated to have almost twice the vitamin content of other children's cereals, so you can guess where that led the commercials. The Kaboom Clown cavorted around his circus's three rings, feeding the cereal to the most incompetent performers. With a thunderclap sounding amazingly like "KABOOM," it transformed a skinny weakling into the circus strongman and helped acrobats and bareback riders (both male and female) rid themselves of what could have become terminal clumsiness. The Kaboom Clown did not have much of a personality of his own, but in his own way, he helped General Mills close out the decade with a big bang.

Kellogg's Best to You

The 1960s

When we last left Kellogg's holding down the fort in Battle Creek, the company was just finding its biggest television advertising success to date with the Hanna-Barbera cartoon shows. That success, not to mention the Hanna-Barbera studio itself, would continue to grow as the 1950s turned into the 1960s.

One of the new cereals introduced at the beginning of the decade was All Stars. Since the tiny star-shaped oat rings were sugarcoated, it seems safe to say they were Kellogg's answer to General Mills' recently introduced Frosty O's. For a mascot character, the Leo Burnett agency materialized the Wizard of Oats. In his tuxedo and top hat, the white-haired necromancer danced around in his animated commercials to a jingle that was as close to MGM's "We're Off to See the Wizard" as copyright laws would allow. The Wizard did not stick around for long before going back behind his curtain, but All Stars would become a Kellogg's mainstay of the decade, changing names and mascots every few years.

Humbug and genuine wizards aside, the Hanna-Barbera shows were the true magic workers when it came to Kellogg's promotion. By 1960, the cereal boxes were encouraging kids to join the "Huck Hound Club,"

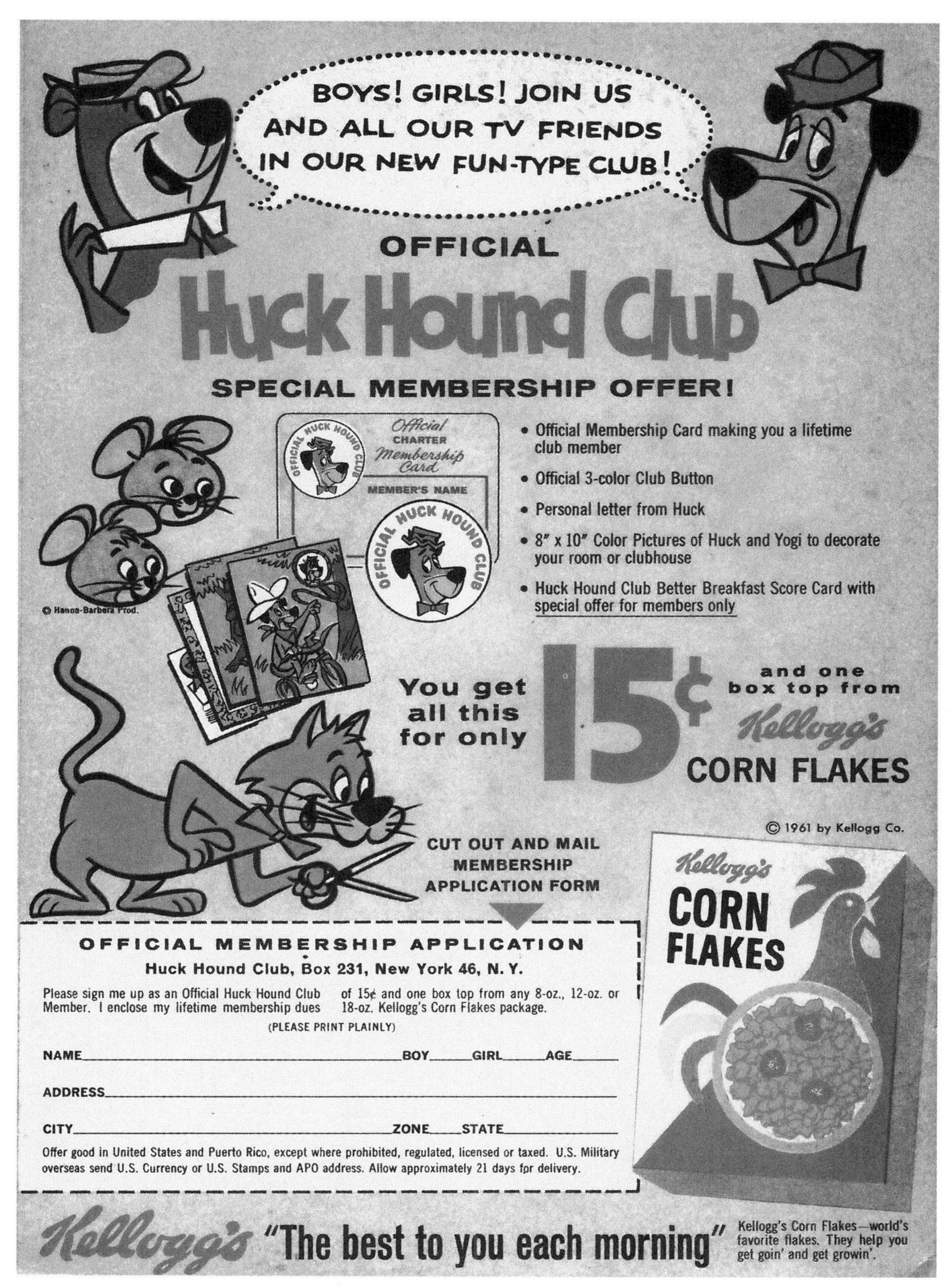

Among the many tie-ins between Kellogg's and the Hanna-Barbera cartoons was the Huck Hound Club. With lifetime dues of only fifteen cents, how could anyone go wrong? A complete set of the club's materials today might set you back a hundred bucks.

and Kellogg's issued long-playing record albums containing sound tracks from the *Huckleberry Hound* and *Quick Draw McGraw* cartoons.

In the midst of all of this, Cocoa Krispies changed mascot characters, but the new one had about as brief a life as the Wizard of Oats. Jose the Monkey was put back on a banana boat headed home while his replacement was Coco the Elephant. Unlike General Mills' Twinkles, who was an orange elephant often misidentified as pink by latter-day historians, Coco really was a pink elephant—although what that fact had to do with chocolate-flavored cereal was never made clear. At the same time that he lumbered onto the boxes, Cocoa Krispies' flavor was changed from bittersweet to milk chocolate, but Coco would have little chance to enjoy it. Something even bigger than a pink elephant was about to take place.

First, in January 1961 Kellogg's began sponsoring a new Hanna-Barbera show starring an old friend. Yogi Bear had proven to be the breakout star of *Huckleberry Hound*, outclassing even the title character. His reward was *The Yogi Bear Show*, each episode of which featured one adventure of Yogi, one cartoon starring Yakky Doodle the talkative duck, and a third segment with the only possible threat to Yogi's star status, the would-be actor Snagglepuss the lion ("Heavens to Murgatroyd!").

As with the earlier shows, the *Yogi Bear* opening theme song managed to incorporate the sponsor into the action, with Yogi driving the ranger's stolen jeep into a Kellogg's billboard. This, however, was not the biggest connection between character and sponsor. With the Hanna-Barbera characters dominating the children's programming ratings, Kellogg's took a revolutionary step that would rarely be repeated: the existing mascot characters for OK's (Big Otis), Cocoa Krispies (Coco the Elephant), All Stars (the Wizard of Oats), and Sugar Smacks (Smaxey the Seal) were all shoved off the breakfast table and into the kitchen trash can, and their box-front positions immediately filled with the Hanna-Barbera TV stars. (One suspects only the long-established tenure of Tony the Tiger and Snap!, Crackle!, and Pop! prevented those icons from being given a similar unceremonious dumping. Somehow Sugar Pops Pete managed to hang on as well.)

There was at least some attempt to maintain continuity between the OK's box with Big Otis and the new design featuring Yogi Bear. Initially, Yogi appeared in a kilt and was shown flexing a huge bicep, just as Big Otis had previously been depicted. (As with Woody Woodpecker's commercials for Rice Krispies a few years earlier, the gluttonous Yogi would not be most people's first thought as the epitome of strength and fitness.) In the commercials, Yogi commented on the change: "Note the

This unusual publicity still depicts the transfer of power from Jose the Monkey to Coco the Elephant as spokesman for Cocoa Krispies. On the single-serving boxes in Coco's hand, notice the Wizard of Oats on All Stars and Quick Draw McGraw on Sugar Smacks.

improvement? They put a bear on the box, which improved it so much they decided to improve the cereal too!" Bantering with the off-screen announcer, Yogi summed up OK's as "The one with the bear on the box—suitable for framin'!"

All Stars had had hardly any time to establish the Wizard of Oats, so he was barely missed. His spot was filled by Huckleberry Hound in a baseball cap, going along with the All Stars name. A year later, the name was changed to Sugar Stars, but Huck and his all-star uniform remained the logo.

Kellogg's other two redesign jobs would last through most of the rest of the decade. Sugar Smacks had sent Smaxey the Seal back to the aquarium in favor of Quick Draw McGraw, holding up a bowl of the cereal on the front of the box. Some of these commercials took a considerably different approach from the other Hanna-Barbera ads. One featured no

Huckleberry Hound replaced the Wizard of Oats as the mascot for All Stars, which soon changed their name to Sugar Stars.

animation at all, only an invisible pencil slowly drawing a scene of Quick Draw and his sidekick, Baba Looey, at the breakfast table as the voices of an off-screen father and young daughter commented on the action:

DAD: Well, what'll it be this time?
GIRL: The one about the funny sheriff.
DAD: All right . . . is this his horse?
GIRL: (giggles) No, that's the sheriff! He's Quick Draw McGraw!
DAD: Oh yeah, he's the fellow on the Sugar Smacks packages.

The longest-lasting Kellogg's/Hanna-Barbera box was the Cocoa Krispies design that replaced Coco the pink elephant with Snagglepuss. The writers seem to have had a particularly fun time crafting the Snagglepuss commercials, as the ham-actor lion's delivery naturally lent itself to hilarious dialogue. In one of the early spots, Snag does what no other cereal mascot has ever done and explains how he "got the job of balancin' a bowl of Cocoa Krispies" on the front of the box. We see him practicing his balancing ability at the circus and in a parody of the famous MGM lion logo ("Roar, even"), until he is discovered by the Kellogg's people. "So here I am, balancin' a bowl of Cocoa Krispies on my pinky," he says. "I hardly ever drop 'em, hardly . . . (CRASH!) . . . ever."

In another episode, Snag narrates the tale of how he woke up one morning and discovered that everything he touched turned to chocolate (a la King Midas). "I didn't know what caused it, even," he states, cueing the appearance of a magical fairy godmother with June Foray's standard "granny" voice.

"But I do, Snag ol' Puss," she cackles. "I've given you the chocolate touch . . . whatever you touch, you can chocolatize."

"So I lived like a king," Snag continues, "with a chocolatized throne in a chocolatized castle. I had a chocolatized motorcycle, even . . . but it wasn't practical."

After concluding that only certain things should taste like chocolate, namely Cocoa Krispies, for a finish Snagglepuss accidentally touches the fairy godmother ("Hey, not *me* too!") and turns her into a miniature chocolate statue.

Speaking of Snagglepuss's vocal abilities, it should be mentioned here that all four of the new Kellogg's spokes-characters (Snag, Huck, Yogi, and Quick Draw) were played by the incomparable voice actor Daws Butler. In the case of three of those roles, Butler's performance would have gone right past the audience, enjoyable but with most people never giving any thought as to the talent behind them. With Snagglepuss, things took

Just as Jose the Monkey had to give up his job to Coco, the pink elephant had to bow out when Snagglepuss became the Cocoa Krispies trademark. Daws Butler's delivery as the voice of the hammy lion made Snagglepuss one of the most memorable characters Kellogg's ever had.

a hairy turn. That voice, at least in the beginning, was played by Butler as a loose impersonation of veteran comedian Bert Lahr, specifically Lahr's role as the Cowardly Lion in *The Wizard of Oz*. Celebrity imitations were nothing new in cartoons, but in this case, Lahr was a bit, shall we say, a-lahrmed. (Heavens to hokey jokes!) You see, at the time of Kellogg's new campaign, Lahr was raking in some serious cash as spokesman for various products, especially Lay's Potato Chips ("You can't eat just one"). He reasoned if ad agencies thought he was also pushing Cocoa Krispies as the voice of Snagglepuss, it might hinder his being hired for other jobs. Therefore, the closing shot of each of the Snag/Krispies commercials featured a subtitle at the bottom of the screen: "Snagglepuss Voice by Daws Butler." Not only did this mean Butler became the only voice actor to ever get a credit line in commercials, but it meant his name was exposed to audiences over and over again—something about which most anonymous cartoon voice actors could only fantasize.

Butler's myriad voices could also be heard when the entire Hanna-Barbera cast appeared in commercials for cereals on which they were not featured as box-front mascots. Everyone from Huck to Yogi, and many of their supporting players, pushed Corn Flakes; as Huck pointed out in one commercial, "No one, not even Kellogg's, has ever been able to make another cereal people like so much!" In a different sort of plot, Yogi fakes amnesia in order to trick Ranger Smith into giving him a bowl of Corn Flakes, since he could *never* forget their distinctive taste. Butler's voices were at their funniest when mice Pixie and Dixie and beatnik cat Mr. Jinks were featured in a Raisin Bran commercial spoofing the then-current age of live TV. The "meeces" are responsible for turning the teleprompter, but their varying speed causes Jinks to slow down or quicken his delivery accordingly, until the whole commercial (and Jinks's sanity) is in shambles. "So, folks, have a ball . . . uh, I mean a bowl . . . uh, a bowling ball. . . ."

Regardless of Butler's sterling reputation, when it came time to introduce a new character, and a new cereal, in 1963 the Leo Burnett agency enlisted the biggest name in the cartoon voice business. Mel Blanc would be the first voice of Toucan Sam, the colorful mascot for Kellogg's newest product, Froot Loops. (In case you have ever wondered, the name was spelled that way not only because it looked better, but so it could be a trademark. "Fruit Loops" would have been disqualified for being merely descriptive. Kellogg's had fought and lost a battle with other cereal companies over the name Raisin Bran back in the 1940s and was not about to let that happen again.)

In retrospect, the choice of Blanc to play Toucan Sam was somewhat odd, as the role called for none of his legendary skill at characterization. Toucan Sam had what basically amounted to Blanc's normal speaking voice. He interacted with two young toucans, garbed in diapers and baby bonnets, who much later would be designated as his nephews in the Huey, Dewey, and Louie tradition. As with real birds, their gender was difficult to pinpoint in their initial appearances, although their voices were supplied by two of the busiest female voice actors in the business, Robie Lester and Nancy Wible.

Toucan Sam stuffed toy

50¢ and one Kellogg's Froot Loops box top

Tastes like fruit—goes "crunch" to boot!

Crispy little cereal circles made from oats. Zinged up with real orange, lemon, and cherry flavors. Sparkled up with big, crunchy sugar crystals. Charged up with important oat protein. At your grocer's now.

Mail coupon for your Toucan toy

SAM TOUCAN TOY
BOX 538, REIDSVILLE, NORTH CAROLINA 27320

Please send me _____ Stuffed Toucan Toy(s). I enclose 50¢ in coin (no stamps, please) and one box top from Kellogg's Froot Loops for each Toucan Toy ordered.

(PLEASE PRINT PLAINLY)

NAME________________________

ADDRESS________________________

CITY____________STATE________ZIP NO.________

Offer good in the United States and Puerto Rico, except where prohibited, licensed or taxed. Make check or money order payable to Toucan Toy. Allow approximately 28 days for delivery.

The year after his introduction, Toucan Sam was offered as a stuffed doll. The toy shared the same mile-long beak as the artwork on the box, but lacked the "real" Toucan Sam's tall hat bedecked with tropical fruit.

The whole gimmick for the Toucan Sam commercials was that he gave a good portion of the spiel in pig Latin:

SAM: Listen, have you noticed something very special about new Kellogg's Oot-fray Oops-lay? They ell-smay so elicious-day!
NEPHEW 1: Yeah, they smell so good too.
SAM: Smart kid.
NEPHEW 2: They don't smell any different to me.
SAM: Why not?!
NEPHEW 2: I got a cold . . . ACHOO!

As with the earlier introduction of Sugar Pops Pete, there was only a slight resemblance between the animated Toucan Sam and his more frequently seen image on the box. The printed Sam was a much more flat 1960s graphic design, with a beak that stretched most of the width of the cereal box. He also wore a tall hat stacked high with the lemons, oranges, and cherries from which Froot Loops got its flavors. The animated Sam ditched the hat (probably because it would have been too complicated to animate frame-by-frame), and his beak was shortened to allow for normal mouth movement.

Even while new characters such as Toucan Sam were being introduced, Kellogg's had not forgotten about its veteran legends. During the early 1960s, Tony the Tiger was teamed with his son Tony Jr. (voiced by Hal Smith) for a series of humorous commercials ranging far from the striped feline's original, more realistic characterization. In one of them, Tony sings a jingle with Thurl Ravenscroft's deepest bass tones, while Tony Jr. encourages the audience to sing along, Mitch Miller style: "Get on the beam, put a tiger on your team/With Kellogg's Sugar Frosted Flakes, grrrrrr-eat!" Tony Jr. is upset because the viewers won't sing: "They just sit there eating Sugar Frosted Flakes!" "We *want* 'em to, Son," chortles Tony. Apparently the cub is unfamiliar with the purposes of advertising: "Well, that's a funny way to learn a song!"

Other adventures combined the animated Tony and Junior with live-action scenes via relatively sophisticated technology predating *Who Framed Roger Rabbit* by twenty years. For example, the two Tonys are seen at a rodeo, where the cowpoke in charge is unimpressed with Papa Tony's resumé. "We'll see how grrreat you are," he mumbles. Ready to ride out of the chute, Tony becomes alarmed when he hears the announcer boom that he is mounted on the back of a Brahma bull. "BRAHMA? I thought they said MAMA!" yelps Tony, just before the bull gives him a wild ride all over the arena.

The Rice Krispies commercials with Snap!, Crackle!, and Pop! also grew funnier in the early 1960s. By now the trio had a memorable jingle sung in counterpoint—i.e., each elf sang his own version simultaneously, blending into an irresistible harmony. The end result was something like this:

Snap! What a happy sound!
You gotta have Crackle! or the clock's not wound;
You can't start hoppin' 'til the cereal's Poppin'!
Snap! Crackle! Pop! makes the world go 'round!

The miniature music makers would engage in all manner of tomfoolery during these productions. In one of the more unusual renditions, each finds himself being injured by some inconsiderate person slamming the cereal box down on him, or being knocked in the head by one of the strawberries being heaped into the bowl. As they sing their big finish, Snap!, Crackle!, and Pop! are bandaged and limping, parodying the famous "Spirit of '76" painting.

Naturally, not all mascots went on to have the longevity of the Krispies gang, or Tony, or even Toucan Sam. There were some who hung around only for the brief time their cereal was on the market and then disappeared into obscurity. Such was the case of a certain unnamed blue giraffe who appeared on boxes of Triple Snack, a Kellogg's concoction made up of Sugar Pops and Sugar Smacks mixed together with roasted peanuts. It would take some serious marketing analysis to determine just why this mixture failed to resonate with shoppers, but it could not be blamed on the catchy Triple Snack jingle performed in the commercials by the giraffe and an also unnamed little boy. It resembled the latest Rice Krispies song, with the boy and the giraffe singing their lines simultaneously, but it ended with a typically 1960s psychological spin:

BOY: I didn't know giraffes could sing.
GIRAFFE: I don't think of myself as a giraffe, especially.

Out of all of these new Kellogg's/Leo Burnett campaigns of the 1960s, one traditional theme seemed to be missing: the old "super strength" routine. (Yogi Bear's flex on the OK's boxes hardly counted, as it only stemmed from Big Otis's earlier pose.) Someone must have realized they were about to let a long-established tradition die out, so in 1964 the next new cereal brought it back in a big way.

Apple Jacks were almost indistinguishable from Froot Loops in appearance, but their cinnamon coating and bits of real apple sticking to the little rings gave them a different sort of flavor. Their mascot, Apple

Triple Snack was not one of Kellogg's most successful products, but its blue giraffe mascot sang duets in the commercials with this young boy wearing a pith helmet.

Jack (who else?), was a lively apple with stick-figure arms and legs, voiced by the third big name in the cartoon acting field, Paul Frees. The initial Apple Jacks campaign was built around the slogan, "An apple a day keeps the bullies away," which seemed as good a reason as any to resurrect the Popeye format.

In one early commercial, a small boy is about to go swimming but finds the pool already occupied by three evil kids whose parents probably encourage them to go play in the street. They taunt the new arrival with such bon mots as "Hey, lookit da tadpole!" and "Where's yer water wings?" until the poor kid dejectedly leaves and bumps into Apple Jack. "Hey, sonny, you better load up on Kellogg's Apple Jacks!" the fruit-headed huckster advises, doing a tap dance to the "keeps the bullies away" jingle while the youngster chows down. It would have been too much like General Mills' idea if the boy grew a bicep with an Apple Jack ring displayed on it, so instead, his chest merely expands while he gets a tough look on his face. Roaring, "BULLIES!" he swims over to the diving board on which the unholy trio is perched, bends it downward with his bare hands, and shoots the meanies away into the sky as if it were a catapult. Frees, as Apple Jack, delivers perhaps the best line of the whole commercial at the end: "Apple Jacks will not be sold to bullies! They're available in Snack Packs, too—uh, Apple Jacks, not bullies!"

In a similar story line, Apple Jack helps the same put-upon kid (or at least one who looks exactly like him) when the bullies grab his two-gallon cowboy hat and toss it high into a tree. He consumes the cereal, squeezes the bullies' baseball in his hand until it explodes, and soon has the brats at his mercy. They not only retrieve his hat and place it back on his head, but brush up his clothes and give him a shoeshine too.

The Apple Jack who appeared on the loud green cereal boxes differed from his animated counterpart: he was an actual carved figure, rather than a line drawing, so his arms and legs were invisible in that medium. Kellogg's apparently liked the hand-carved three-dimensional look, because at the same time as Apple Jack's debut, the company also introduced Corn Flakes with Bananas and a similarly created character, Pronto Banana. The cereal went nowhere, even though its big selling point was that it contained dehydrated banana slices that reconstituted themselves when milk was added. Pronto Banana wore a straw boater and used one of his peels as a hand, but had no real personality to speak of. Another carved character, a tuxedo-wearing ice cream cone, bedecked boxes of Kream Krunch. Historian Scott Bruce claims that this mixture, consisting of freeze-dried chunks of vanilla, strawberry, and orange ice cream mixed

The fruit-headed mascots for Apple Jacks and Corn Flakes with Bananas were created by the unusual method of actually having them modeled out of real fruit, rather than depicted as drawings. Apple Jack hung around the tree for several years, but Pronto Banana just did not have enough a-peel.

with a close cousin of Frosty O's, was test-marketed in only three cities before being put back into the deep freeze. The ice cream man never even had a chance to get a name or a personality.

Besides the various Hanna-Barbera shows, Kellogg's next big outlet for its TV commercials was *The Beverly Hillbillies*. There was a natural association between the hillbillies and corn (in more ways than one), so the association turned out to be a very profitable one. That "story of a man named Jed" begat something of a craze for television hillbillies, which is no doubt what led to the next big mascot switch.

Huckleberry Hound, a rural character in his own right, was still appearing on boxes of Sugar Stars in 1965 but he wouldn't be for long. A new, even more hickish hayseed was about to take his place: Hillbilly Goat. Perhaps because Huck had been with the company for so long, Kellogg's decided the changeover deserved more than a simple, silent swap of one box design for another. Huck and Hillbilly Goat appeared in a commercial together (the goat refers to his host as "Cousin Huckleberry," hinting that interspecies inbreeding is rampant in the backwoods country) in which Huck patiently explains the new features of Sugar Stars:

HUCK: Hold on thar! Th' hole in th' middle is GONE.
GOAT: (panicking) Don't ennybody leave th' room! Who took it, Huck?
HUCK: Kellogg's plugged it up!
GOAT: (gasp) They DID?
HUCK: They made Sugar Stars crunchier an' bigger, and give 'em caramel coats!
GOAT: Well, what'll they think of next?

Huck then points out the box has changed, too, although he refrains from specifying that the goat's picture has replaced his own. Upon being told the box is different, Hillbilly Goat takes a couple of bites out of it: "Don't taste enny diff'rent to me!" Fortunately for Kellogg's, when the time came to cast a voice for Hillbilly Goat, they were again able to turn to the Hanna-Barbera staff to find someone besides Daws Butler who had been making a living by doing a hillbilly accent. Howard Morris had already played several Hanna-Barbera characters, in particular Mush Mouse, antagonist to fellow hillbilly Punkin Puss. Even earlier, as most TV fans know, Morris had been making irregular appearances on *The Andy Griffith Show* as the incorrigible Ernest T. Bass, so he simply dragged out the same voice he had used as those characters and implanted it into Hillbilly Goat's throat.

Huckleberry Hound appeared in one final Sugar Stars commercial in 1965, explaining the product to his cousin Hillbilly Goat before turning mascot chores over to the omnivorous hayseed.

Once Huck was no longer necessary for the transition, Hillbilly Goat starred in his own commercials. One of them shows the panic ensuing in a typical rural town when word spreads that he is on his way. This is not because he is a bad guy, but because in true cartoon goat fashion, he eats everything in his path, from front porch columns to milk pails. His appetite is assuaged by Sugar Stars, over which he sings: "Shore do love those crunchy oats/Shaped like stars with caramel coats/Good fer kids 'n' hillbilly goats/Human kids, that is!" The next year, Kellogg's changed the name of the cereal again, this time simply to Stars, but Hillbilly Goat remained the logo character until the end of the decade. He was briefly replaced by a yellow representation of the Man in the Moon wearing a nightcap, but by 1970 or thereabouts, Stars cereal had used up its last wish and disappeared into the black hole of obscurity.

By now it was an accepted fact that an identifiable mascot would help the marketing of any cereal more often than not, but Kellogg's still had one major product being pushed during Saturday-morning television that lacked an animated character. The problem child was Raisin Bran—yes, the same member of the family that had given the parent company trouble twenty years earlier when it was found that the name could not be trademarked. (Inasmuch as it merely described the ingredients, plenty of other companies were able to market their own Raisin Bran, including Skinner's—which was the original—and Post.) A partial solution to Kellogg's dilemma was to assign a character to Raisin Bran, separate from the earlier hilarious commercials with Pixie, Dixie, and Mr. Jinks.

Since the sun was responsible for turning grapes into raisins, someone had the bright idea of making the sun the new Raisin Bran logo. But how could anyone be expected to impart personality to such an impersonal force of nature? One solution was to give the sun a voice by Daws Butler and the next was to have the orb engage in give-and-take banter with the grapes and raisins, which would be played by Robie Lester. The somewhat shy sun, in a desperate effort to be liked, would try to join the raisins in the cereal bowl, claiming first, "I'm a grape!" or, "I'm a raisin!" or, "I'm a flake!" while the raisins would correct him: "No, you're the SUN!" It all sounds a bit flaky, but the sun's simplified, smiling face became the permanent logo for Raisin Bran—or at least the Kellogg's version of that generic product.

As we saw earlier, Kellogg's had long held the belief that Corn Flakes needed no such character, apart from having various guest stars tout its many good qualities. Cornelius the Rooster had been created especially to tie in with the *Huckleberry Hound* opening and closing sequences but remained an almost abstract piece of artwork on the box. The box-front rendition would not be changing, but by 1966, Cornelius (or Corny) would be seen on TV in a much more full-fledged manner.

It does not take very close examination of the new Cornelius commercials to see that they were certainly the most lavish animation ever produced by Hanna-Barbera. One can easily imagine the studio's artists wishing they had such budgets to work with on all of their programming. As we saw in the case of General Mills and Jay Ward, though, the ad agencies were willing to spend a whole lot more cash on the commercials than they were on what was supposed to be an entertaining TV show.

Whereas the early 1960s Frosted Flakes spots had put the animated Tony into the real world, the Corn Flakes campaign put live actors into a cartoon world. When it came time to determine just how a talking rooster would sound, Kellogg's reached back into its own past and cast the unmistakable Andy Devine as the voice of Cornelius. The other voice heard regularly in the commercials belonged to Dick Beals, who enjoyed a long and varied career in radio and animation but is best remembered as the voice of Speedy Alka-Seltzer. In each commercial, Corny would interact with various live-action boys, but no matter how the young actors looked, Dick Beals was always their dubbed voice.

For Kellogg's ninetieth anniversary in 1995, the original simple Raisin Bran sun design was brought out of retirement to once again grace the fronts of special commemorative boxes.

The basic plot of each commercial was that during breakfast, the boy would wish to go on some particular adventure and make a magic gesture toward the abstract rooster on the Corn Flakes box. Corny would then spring to life, calling his young friend "boss," and off they would go with Corny making a constant stream of wisecracks about their predicaments. In one of these, the boy wishes he could discover America with Christopher Columbus, and soon he and Corny are aboard the *Niña, Pinta,* or *Santa Maria.* The boy announces that he doesn't see Columbus, while a seasick Corny replies, "Maybe you'd rather see Columbus, Ohio." Finally, Columbus shouts at them from the crow's nest that land is imminent, but the boy can't see it. "I guess I'll have to tall up," he shrugs, cueing the commercials' other consistent scene: Corny gives him a bowl of Corn Flakes, causing the boy to grow to monumental heights. (Or, since the action is taking place in the box-front rooster's world, maybe the boy simply becomes normal human size.) With his heightened altitude, the boy is able to discover America on the distant horizon.

In another adventure, the boy/rooster duo goes into the African jungle, where the youngster has to "tall up" to rescue them from a lion. Once the kid is big enough to pet the purring lion like a housecat, Corny risks trademark infringement by lifting a line from another cartoon studio: "I tink I tee a puddy tat!" On another occasion, they visit prehistoric days so the boy can get his wish to ride a dinosaur. When their brontosaurus transportation encounters a fierce tyrannosaurus, Corny gulps, "Uh, oh, here come the folks from next door!" Upon making their escape, the rooster makes another corny joke: "That dino's really SORE!"

Devine's distinctive delivery of his funny dialogue was certainly a highlight of every spot, and Dick Beals always sounded more authentic as a young boy than any of the kids' real voices would have. It seems Hanna-Barbera learned a thing or two about combining live actors with animated backgrounds and characters from doing these commercials, because the next year the studio produced a well-received *Jack and the Beanstalk* special following the same format, and in 1968 introduced a prime-time series, *The New Adventures of Huck Finn*, in which the child actors spent most of their screen time in a cartoon world.

Meanwhile, back in Battle Creek, another cereal was getting ready to replace its Hanna-Barbera mascot with something more original. Quick Draw McGraw was still on the front of the Sugar Smacks boxes when a commercial introduced the punchy Smackin' Brothers in 1966. Paul Frees served as narrator, while the two battling brothers were portrayed vocally by Robie Lester and Gloria Wood. We have already seen how commercials

Sugar Smacks' Smackin' Brothers were certainly two of the most pugilistic characters to ever advertise cereal. Is it still considered child abuse when children abuse each other?

were far from immune to animated cartoon violence for comedy's sake, but the Smackin' Brothers took it to a different level. Since they were actual human beings—albeit cartoony ones with oversized heads and tiny bodies, forever attired in boxing gloves and trunks—their constant whacking of each other could be a bit painful to watch. When the Gloria Wood brother would spy his sibling munching on Sugar Smacks and sing, "Hey, gimme a smack," the Robie Lester brother was a little too willing to oblige. "And 'cause you're me brudder, I'll give ya anudder," he would threaten his prone twin. When the announcer would remind them they were talking about Sugar Smacks, Lester would belt out a rapid-fire patter song:

> He means with honey on the outside, sugar on the inside,
> Puffs of wheat, smackin' sweet,
> A taste that knocks you off your feet,
> Kellogg's Sugar Smacks!

Once the Smackin' Brothers replaced Quick Draw on the box, only Snagglepuss was left still holding down stage left on Cocoa Krispies. The transition to a new mascot took a bit longer and was a little more drawn-out in the case of Snag's replacement. The first step was to co-star the easygoing lion with a new character named Koko, who could be interpreted as either a caveman or a parody of Tarzan. (Either way, Koko was only about two feet tall.) Koko's voice was supplied by perhaps the only actor who could reasonably give Thurl Ravenscroft some competition in the deep-bass department: Ted Cassidy, best known as butler Lurch on *The Addams Family*. Cassidy was loaning his growl to several Hanna-Barbera shows of the period, no doubt explaining how he happened to be associated with Snagglepuss. Eventually Koko starred in the commercials by himself, but Snagglepuss remained on the front of the box until another character could replace him—and this time there was no doubt the new Cocoa Krispies spokesperson was indeed a caveman.

The caveman's name was Ogg, and his infrequently seen wife was Kell. Beginning to get the picture? The commercials with Ogg were as unique in their own way as the live action/animation Corn Flakes adventures: Ogg was created via stop-motion animation, and as we saw way back with the three Sugar Crisp bears in the 1950s, that medium was usually too expensive and time-consuming for cereal advertising. Reportedly Ogg's voice was supplied by Paul Frees (it figures), and his career as Cocoa Krispies mascot would extend well into the next decade.

(Frees, by then, had also taken over the role of Toucan Sam from Mel Blanc. Frees's Sam was a vocal impersonation of dignified actor Ronald Colman, but unfortunately there do not seem to be any surviving commercials from that transition period to see just how such a drastic change in voice and personality was handled.)

Ogg made his debut in 1968, and over the course of the next year, there would be further changes in Kellogg's lineup. The world of 1969 was almost a different planet from the world of 1960, so it was only natural that some of the remaining vestiges of the past would be jettisoned. Sugar Pops Pete had been faithfully carrying out his duties since the late 1950s, but in 1969, he too was kicked down a prairie dog hole. Sugar Pops stuck with the western theme it had employed since the beginning, however, and Pete's replacement was a live actor billed as the Whippersnapper. As a character, the Whippersnapper had about as much personality as the Cream of Wheat chef. He did not even speak, his only function being to crack his huge bullwhip and transform corn on the cob into Sugar Pops. Narration was handled by the familiar western tones of Rex Allen.

Two other cereals introduced at the end of the 1960s took their approach about as far from the westerns as could be imagined. Kombos was an unsuccessful attempt at marketing chocolate, orange, and

Toucan Sam received cosmetic surgery in 1969, emerging with a shorter beak and a slightly different color scheme. His original, lengthy bill contained one black section, one orange section, one yellow section, and two pink sections, but his new design cut those down to one segment of each color.

strawberry-flavored Corn Flakes. The spokes-character for all three flavors was the Blue Gnu, a bucktoothed beast wearing a blue peacoat, but neither he nor Kombos had what it took to survive in the cereal-aisle jungle. Slightly better remembered was Puffa Puffa Rice, which did basically the same thing to rice that Sugar Smacks did to wheat. Puffa Puffa Rice was sold with a Hawaiian theme, first with no character on the box except a colorful volcano blowing pieces of the cereal out of its crater. Soon afterward, the faceless volcano was replaced by two Hawaiian kids paddling an outrigger canoe, probably inspired by the title sequence of the hit police drama *Hawaii Five-O.*

Just before the turbulent 1960s gave way to the 1970s, another character took his last bow. Apple Jack must have grown tired of helping kids beat up neighborhood bullies, because he left the boxes and was replaced by a simple drawing of an apple-shaped car, complete with steering wheel. Apple Jack continued to appear on the single-serving boxes for many more years. In fact, the "snack pack"–size Kellogg's packaging during the 1970s often served as a reminder of how things used to be. On the full-size Froot Loops boxes, Toucan Sam was given cosmetic surgery during 1969, emerging with a considerably shorter beak and different color scheme. On the single-serving boxes, though, he could still be seen with his original beak and fruit-decorated hat for almost another ten years. The snack-pack boxes became miniature time capsules, preserving nostalgic memories from a decade that practically defined changing times.

Start Your Day a Little Bit Better

Post in the 1960s

At the beginning of the 1960s, Post found itself lagging behind the competition when it came to kids' marketing. Kellogg's had hooked up with Hanna-Barbera, while General Mills was trying to pull a rabbit out of its hat with *Rocky and His Friends*, and Post was simply drooling to get a bigger piece of the cartoon action. With its sponsorship of *Mighty Mouse Playhouse* wearing out after so many years, Post took no chances and aligned itself with one of the biggest stars Toontown had ever produced.

A large portion of the Warner Bros. cartoon library had been syndicated to local TV stations across the country since 1956. However, another, newer part of the collection was held back until the right vehicle came along. In October 1960, the ABC network premiered *The Bugs Bunny Show*, a weekly prime-time romp with the wascally wabbit and all of his looney pals.

Post followed both Hanna-Barbera's and General Mills' lead and had the Warner Bros. cast pushing practically every brand of its cereal in the commercials—however, the company drew the line at having them appear on the fronts of the boxes as logos. Only one of them was ever afforded such an honor: "that Oscar-winning rabbit Bugs Bunny" briefly served as the mascot for an obscure Post concoction known as Top 3, a mixture of corn flakes, wheat flakes, and puffed rice. Otherwise, the cartoon crowd stuck to games and prizes on the box backs, and of course their weekly hilarious commercials.

Just as with so many other characters conscripted in the cause of selling cereal, it seems odd that Bugs Bunny (who always outwitted his opponents with superior intellect) would be used to demonstrate the added strength one got from Sugar Crisp, Alpha-Bits, or any of the others. On the *Bugs Bunny Show*'s familiar stage setting, the jealous Daffy Duck would continue his attempts to become sole host by aiming a huge mallet at Bugs's cranium, only to have it crack when striking the rabbit's fortified fist or else bounce off his ears and drive Daffy into the ground. Bugs would pontificate on the benefits of "muscles of wheat," to which Daffy would mutter to the audience, "Muscles in his ears, yet."

Most of the commercials involved Bugs playing more true to form and tricking Elmer Fudd or Yosemite Sam out of their cereal (a theme Post would come to know well in future years). Periodically, the characters would sing and dance to Post's newest jingle, "Start your day a little bit better/Start your day with the cereals from Post." But, when the *Bugs Bunny Show*'s exceptionally large cast was not doing the sales pitch

Although Bugs Bunny was better known for outwitting his foes with superior intellect, during Post's sponsorship of ABC-TV's *Bugs Bunny Show* the scwewy wabbit demonstrated the "muscles of wheat" one could obtain by eating Sugar Crisp.

So-Hi, an ethnic caricature of a young Chinese boy, was the mascot for Post's Frosted Rice Krinkles throughout the 1960s.

themselves, Post was introducing a whole new repertory company of its own, which would come to dominate the company's advertising for the rest of the decade.

One of the first was the new mascot for Rice Krinkles. Since rice is traditionally associated with the Orient, the spokesman for Krinkles was So-Hi, a Chinese boy with a bowl haircut and another type of bowl—filled with the cereal—he wore as a hat. With his dialect habit of substituting *r*'s for *l*'s, one could reasonably make the argument that So-Hi fit all the criteria of an ethnic stereotype. As with the Cream of Wheat chef in the olden days, though, So-Hi was not a malicious caricature; in fact, the worst that can be said of him is that he displayed no real personality traits at all beyond his voice (supplied by Robert McFadden) and his knack for ringing a gong every time he prefaced wise advice with the phrase "So-Hi say . . ." Still, he was probably the least memorable of the 1960s Post characters.

In one of his commercials from the early 1960s, the Sugar Crisp bear gives young Christopher Wheat some sage advice on being a hero. The original other two bears were gone for good by this time.

By the time *The Bugs Bunny Show* hopped into prime-time TV, Sugar Crisp had finally achieved its goal of reducing the bears on the box from three to one. That remaining ursine spokes-animal still did not have a name and, in his newest series of commercials, hardly participated in the action at all. One might be tempted to think of them as live-action counterparts to General Mills' Cheerios Kid campaign, starring a new preadolescent hero known as Christopher Wheat. This boy seemed to have an unusual proclivity for being present when trouble broke out, whether it was a bank robbery, gorillas escaping from the zoo, or, in an even more exotic setting, pirates capturing an adorable little girl. Christopher would sadly wish he could help, whereupon the animated bear would come to life and speak to him from the front of the Sugar Crisp box. After being reminded that he has powers far beyond those of ordinary mortals because he eats Sugar Crisp, Christopher would spring to the rescue in sped-up action strongly resembling animated cartoon gags; in less time than it takes to tell, he would have the robbers or gorillas or pirates or other baddies beaten into submission and would receive a kiss of thanks from the girl (proving Sugar Crisp helped his good sense too, unlike the ever-platonic Cheerios Kid).

The plot of the Christopher Wheat commercials was also played out with another live-action boy identified only as "the Big Kid." He was indeed taller than the other youngsters who appeared along with him, but he still required advice from the Sugar Crisp bear to catch on to how to effectively deal with the bedlam that was breaking out all around him.

One major Post cereal was still adrift without a character to anchor it, and that was Alpha-Bits. The ad agency tried to come up with someone who could conceivably be tied to the theme; since Alpha-Bits were shaped like letters, who delivers letters? A mailman, right? Perish the thought! Alpha-Bits could only be delivered by . . . a *POST*man. But how should this faithful courier look and sound when out on his appointed rounds?

Somehow Post ended up with insult-comic Jack E. Leonard, and for one of the few times in cereal advertising history, the character was actually designed to look like his voice actor. Leonard's fast-talking style was entertaining to hear, but his commercials as the Alpha-Bits Postman were too short to really give him a chance to do his best work. In each spot he would deliver Alpha-Bits to an "Our Gang" brand of neighborhood kids, only to run afoul of the group's dog. "I'm not the kind of Postman you fellows chase," he would try to alibi. "All my letters are Post Alpha-Bits!" In one of the funniest variations on the theme, the Postman shows the kids how to arrange Alpha-Bits to spell words of their choosing, and one boy points out that their dog uses the cereal pieces to communicate. "Go ahead, Fido . . . say something in Alpha-Bits," challenges the Postman, looking at a table on which the dog has spelled GRRR. "Very funny . . . HELP!" yells Leonard as the dog chases him up a nearby lamppost.

A character from the past emerged from retirement in 1962 to help with Post's newest product, and that character would turn out to be to Post what Tony the Tiger was to Kellogg's. The comparison was more than accurate because the name (and voice) from the past was unemployed Heart of Oats huckster Linus the Lionhearted, now with his name shortened to simply Linus the Lion. He had also been redesigned, although since everyone knows how a lion is supposed to look, it was not a drastic process. The new Linus no longer had the body of an athlete with an ego to match; in fact, outside of uprooting a tree to demonstrate "the strength of a lion," he was something of a bumbler. Sheldon Leonard continued to supply his

The first Alpha-Bits Postman was both voiced by and designed to look like famed insult-comic Jack E. Leonard. His tenure as a mascot was short-lived, but the concept of a postman who delivered Alpha-Bits' "letters" would stick around for a while.

Unemployed Heart of Oats salesman Linus the Lionhearted came out of retirement with a shortened name, Linus the Lion, to become the much-loved mascot for Post's new Crispy Critters. Unlike in his former life, this Linus was hardly athletic and, in fact, was usually somewhat clumsy.

voice, using his unique delivery to make even the most mundane words sound comical ("delicious" became "dee-li-shee-us" in Leonardspeak).

What was the new cereal warranting so much work? It was Crispy Critters, a sugar-frosted version of Heart of Oats, only shaped like various wild and domestic beasts. A bowl of Crispy Critters more or less resembled a pile of animal crackers. The first couple of years of Linus's commercials consisted of one shtick only: whenever he would speak the name of the cereal, he would immediately be trampled into the ground by a stampeding herd of "the one and only cereal that comes in the shape of animals," as the jingle would burst forth. It mattered not where Linus thought he was safe—in the top of a palm tree, in a submarine at the bottom of the river—the thundering herd would find him every time. He would sometimes succeed in refraining from saying "Crispy Critters" himself, only to have another animal standing next to him blurt out the phrase. Naturally, Linus would get trampled, but his offending companion would not.

The Linus who appeared on the Crispy Critters boxes was slightly different from the animated commercials' Linus. The first box art showed him inside a cage at a circus or zoo, licking his lips over a giant bowl of the cereal. The next year, he was given a top hat and ringmaster's outfit, holding a hoop through which the animal-shaped oat pieces were jumping into the bowl. Finally, around 1964, Linus's two varying images converged and remained the same for the rest of his career.

Post's next two characters were something of a letdown after the slapstick humor of Linus and his friends. Sugar Sparkled Flakes shucked gabby corncob Cornelius W. Sugarcoat and replaced him with a nameless boy genie emerging from a sugar bowl instead of Aladdin's lamp. The genie (who was never given a name) seemed only to emphasize the energy provided by the cereal's sugar coating. Post's Raisin Bran featured the Raisin Counter, a purple-suited caricature of a bespectacled accountant, who guaranteed a bowlful would have a requisite number of ex-grapes.

If one had to choose the year that it all came together for Post, it would have to be 1964. It would be climaxed by possibly the most audacious advertising stunt ever devised to promote cereal, but there were some other elements that had to fall into place. First was the problem of the ironically bland sugar genie on Sugar Sparkled Flakes. The ad agency decided it had been putting the emphasis on the wrong ingredient and instead chose to go back to a corn theme. (Cornelius W. Sugarcoat, however, was not called back to active duty and continued to languish in the unemployment line.) Sugar Sparkled Flakes would have a new mascot—or actually, two mascots. The commercials established Rory Raccoon as a farmer trying to guard his corn crop from the scheming Claudius Crow, the latest in what would prove to be a long line of Post cereal thieves. Rory's voice was supplied by Robert McFadden; Claudius was played by character actor Jesse White, later best known as the lonely Maytag Repairman. Claudius was tireless in his efforts to swipe the corn from under the raccoon's pointed nose.

Meanwhile, Post's contract with Jack E. Leonard had run out, so Alpha-Bits needed a new mascot. Everyone loved the "postman" concept so much that the Leonard postman was simply replaced with a new letter carrier: Lovable Truly, a tall, thin USPS employee with the most hick accent Robert McFadden could coax out of his voice box. The first version of Lovable appearing on Alpha-Bits boxes was so dopey-looking and anorexic that people had to wonder whether he would ever be able to read the addresses on the mail he delivered, much less lug a heavy sack of it from house to house. Therefore, after only a few months, lovable old

Lovable Truly replaced Jack E. Leonard as the Alpha-Bits Postman and remained on the boxes long after his final animated commercial appearance.

Lovable was redesigned to look slightly more masculine, although he still did not possess what anyone would consider an imposing physique.

Neither did Post's remaining new character of 1964, but in his case it proved to be a deceptive appearance. The one leftover bear on the Sugar Crisp boxes was given a makeover to render him less like a teddy bear, and he emerged as Post's longest-lasting superstar, Sugar Bear. In nearly every way, Sugar Bear was deliberately the polar opposite (and no, that didn't mean he was a polar bear) of his nameless ancestor. First, there was the matter of his voice. The former Sugar Crisp bear had spoken with a high-pitched child's voice, but someone thought it would be funny to have Sugar Bear sing and speak in the tones of Bing Crosby. Hiring the real Crosby would have put Post on the road to bankruptcy, even without throwing Bob Hope into the mix, so the call went out for someone who could do a passable Crosby impersonation. Answering the call was nightclub comedian Gerry Matthews, who later admitted, "Bing Crosby was the only impression I knew how to do!" (There is no record of what Crosby thought of his voice being appropriated for this purpose, although we have already seen how Bert Lahr reacted to Snagglepuss.)

The Crosby voice was only one of the characteristics setting Sugar Bear apart from all his rice-, wheat-, oat-, and corn-fed kinfolk. Strolling

In a long series of commercials, Sugar Bear would find multiple ways to con Granny Goodwitch out of her supply of Sugar Crisp.

through life with half-closed eyelids and half-humming, half-mumbling his "Can't get enough of that Sugar Crisp" theme song, Sugar Bear gave the impression of total passivity. This persona camouflaged the fact that he possessed superhuman strength, but he made no big deal about it. If a tree needed to be uprooted or anything else stood in the way between Sugar Bear and his cereal, he would simply handle it with his bare (or bear) hands without even opening his eyes any wider.

In another break from the expected, while he was supposed to be a good guy, Sugar Bear was a habitual sneak thief. The target of his thievery was Granny Goodwitch, who introduced herself in an early commercial appearance as the only human being who lived in the forest. (Even though a witch, she did consider herself a human, apparently.) Granny's voice was supplied by up-and-coming comedian Ruth Buzzi, who would soon be going on to more fame with *Rowan and Martin's Laugh-In*, but in 1964 she was more concerned about safeguarding her ever-present supply of Sugar Crisp from her friendly nemesis.

In every commercial, Sugar Bear would invite himself to Granny's Sugar Crisp stash, and the good-natured old hag would use her magic to try to discourage him. She might turn her house into a ship and sail away in it, or turn it into a haunted house full of terrors to scare the bear away, but thanks to his effortless feats of strength, Sugar Bear was unstoppable. In one extreme example, Granny turns her house into a rocket and blasts off for outer space. Surrounded by inky blackness, she thinks her cereal is safe at last, only to be greeted by the sight of Sugar Bear pedaling a bicycle past her porthole, mumbling, "Hi, Granny . . . small universe." (Apparently eating Sugar Crisp not only creates excess strength, but also enables one to breathe when there is no oxygen.)

Now, with all these characters fixed in the public's collective mind, it was time for Post to make a step never before attempted by any cereal maker, and it would never be repeated again. In the autumn of 1964, a new Saturday-morning cartoon series debuted on NBC: *The Linus the Lionhearted Show*. Yes, while all the characters were still appearing in commercials on other programs, they were given an entire half hour of their own to fill with animated antics.

Each *Linus* episode actually consisted of four short (only four minutes each) cartoons featuring one or another of the Post cereals cast. Between cartoons, along with the obligatory commercials, there would be an additional continuing story that would play itself out during the half hour. Actually, apart from the opening titles, these "bridging" stories were the only time the characters actually interacted with each other in Linus's jungle kingdom, the cartoons themselves ranging far and wide from such a locale.

In the lead cartoon of each half hour, Linus would encounter difficulties in keeping things between the ditches in his domain. His friendly antagonists included Billy Bird (or "Boid," as Leonard pronounced it) and Dinny Kangaroo, both voiced by comedy great Carl Reiner. Although these two characters were not cereal mascots, one gets the impression that Post wanted to make them familiar to audiences in case that necessity ever arose. The second cartoon of the four would alternate between an adventure with Sugar Bear in his enchanted forest and the continuing attempts of Claudius Crow to swipe corn from Rory Raccoon's field. The Claudius/Rory battles of wits were fairly standard cartoon plots, but the Sugar Bear stories took things in a different direction. Granny Goodwitch made regular appearances, but the main magic-worker in that neck of the woods was the unsavory Mervyn the Magician, who constantly schemed to make everyone's life miserable.

The third cartoon was usually built around Lovable Truly, but as possibly the blandest of any of the characters—and with the relatively unadventurous life a postman usually leads—much of the footage in his cartoons was given over to villainous Richard Harry Nearly ("former silent

The Linus the Lionhearted Show **debuted on Saturday morning television in September 1964. To date, it remains the only cartoon series with a cast consisting entirely of cereal mascot characters. In this jigsaw puzzle, Linus and Sugar Bear enjoy the antics of Rory Raccoon and Claudius Crow.**

movie star and part-time dogcatcher") and his attempts to permanently incarcerate friendly mutt Lawrence. Good ol' Lovable would usually end up coming to Lawrence's rescue. The fourth cartoon was the one that would probably prevent these shows from being released on DVD, if Post were inclined to use them that way. In an obvious swipe of *Rocky and His Friends*' "Fractured Fairy Tales," So-Hi would relate a weekly fable of his own, in which all of the characters would speak the same stereotypical Chinese dialect. As an example, take the episode titled "Little Red So-Hi Hood," in which the host narrates over the action: "The woods velly dangerous, but Little Red So-Hi Hood not afraid, because he stupid." (Perhaps afraid suspicious viewers would make a connection between the character and Red China, the story finishes with So-Hi unfurling an American flag and becoming the superhero "Little Red-White-and-Blue So-Hi Hood.") Each story would end with So-Hi delivering a twisted moral, such as, "Time spent wishing for happiness not worth two cents" and, "When wolf is at door, take it on lamb."

Many TV nostalgia fans have commented on a very odd situation regarding the program's opening theme song. It continued the humorous tradition of the Crispy Critters commercials by having Linus trampled by a stampede—only this time, the herd was real jungle animals and not pieces of cereal. But the strangest thing, from an advertising standpoint, is the theme ("Linus is the one who lines up the fun/He's the host of which we boast") was sung by the male quartet known as the MelloMen. Now, if you will recall from many pages ago, the MelloMen was the group in which Thurl Ravenscroft sang bass—so, in effect, he was singing for Post while simultaneously playing Tony the Tiger for Kellogg's. At one point, the lyrics refer to Linus as "the biggest, the greatest, the sweetest, the most," and some people have claimed Ravenscroft was given "the greatest" line as a dig toward Tony's usual slogan. As clever and underhanded as that would have been, Ravenscroft actually sings "the sweetest" instead, while one of the other singers gets "greatest."

There was certainly nothing sneaky about the way Post promoted the show. On Thanksgiving Day 1964, a 50-foot-tall Linus balloon made its Macy's Thanksgiving Day Parade debut, sponsored by the cereal company. There was even a special episode of the *Linus* show in which he and Billy Bird spent the entire bridging story line just trying to inflate the giant figure. (Linus became the first cereal mascot to appear in the parade, although in 1961 General Mills had put a Bullwinkle balloon in the lineup, and in 1965, they would also add Underdog to the mix. As we have

Post began sponsoring a 50-foot-tall Linus the Lionhearted balloon in the Macy's Thanksgiving Day Parade in 1964. A few years earlier, General Mills had footed the bill to insert a giant Bullwinkle balloon in the annual event.

seen, though, both characters' primary fame was outside of their cereal association.)

In 1965, Post offered as a premium a long-playing record album featuring the *Linus* cast. Since vocal talent such as Sheldon Leonard, Carl Reiner, and Jesse White did not come cheap—not even counting the rest of the actors—at least some money was saved by having all the songs on the LP be public-domain melodies, although with new lyrics to tie them

LINUS

12¢

10155-509

Linus THE LIONHEARTED

Linus and Lovable Truly brave the dangers of the jungle to deliver a mysterious package!

Since *Linus the Lionhearted* had a life as a cartoon series apart from the characters' cereal commercials, much merchandise was produced for sale in department stores, variety stores, and toy stores. This comic book was just one of the dozens of items on the market.

to the appropriate characters. "Ta-Rah-Rah-Boom-De-Ay" was reconstituted as "Hoo-Rah-Rah Sugar Bear," with "When Johnny Comes Marching Home" becoming "When Lovable Truly Brings the Mail." Claudius and Rory sang a duet on "Nothing's More Fun Than Eatin' Corn," to the tune of "Little Brown Jug" and "Casey Jones" was rendered as "So-Hi Say." In many ways, listening to this album gives one an even greater appreciation for the voice talent involved, since without the distraction of the animation, it is easier to catch the many nuances present in, for example, Sheldon Leonard's delivery of Linus's lions—uh, lines.

Besides the LP, there was much more Linus merchandise produced for retail sale, completely apart from premiums. Toy stores featured all the usual suspects: jigsaw puzzles, storybooks, lunch boxes, plastic banks, board games, coloring books, stuffed dolls, comic books, and Halloween costumes were only a few of the many items available (and for which collectors pay staggering prices today).

Twenty-six half hours of *Linus the Lionhearted* were produced for the initial 1964–65 TV season, with an additional thirteen shows for 1965–66. Those thirty-nine total episodes continued airing in various time slots, on both Saturday and Sunday mornings, until the beginning of the 1969–70 season. In the meantime, the world of Post's advertising did not stand still, and some developments came along too late to be incorporated into the body of the show—although one could certainly count on seeing them during the commercial breaks.

A new Post cereal was unwrapped in 1965: Honeycomb. The logo character was a cowboy named the Honeycomb Kid, whom some ad executives thought resembled a caricature of President Lyndon B. Johnson but whose infrequent lines of dialogue were delivered in a John Wayne impersonation. More than anything else, the Honeycomb Kid's adventures seemed inspired by such famous western tall-tale heroes as Pecos Bill. When danger threatened, the Kid would swoop to the rescue riding a gigantic golden eagle. In a typical escapade, the squinty-eyed cowpoke puts out a forest fire by squeezing the rain clouds like sponges until their water douses the flames. Some sources claim the intention was to give the Honeycomb Kid his own segment in the *Linus* show's second season, but for one reason or another, this never came about. The Kid was seen alongside the other Post characters in print advertising but not on the merchandise based on the program.

At some point during 1968, Sugar Crisp was reformulated to boost its vitamin content, and the resulting product was given the name Super Sugar Crisp. Gerry Matthews was still channeling Bing Crosby as Sugar

The squinty-eyed Honeycomb Kid, introduced in 1965, did not get his own adventures in the weekly *Linus the Lionhearted* show, but he did join Linus and Sugar Bear for some newspaper ads such as this one.

Bear's voice, but now the commercials took a slightly different approach. You will recall that originally Sugar Bear was presented as supernaturally strong because of his constant diet of Sugar Crisp; in his new commercials, he aped Popeye a little more closely in that he had to consume a bowl of Super Sugar Crisp before he could leap into action.

The Super Sugar Crisp commercials of the latter half of the 1960s fell into two basic categories. In one series, Granny Goodwitch would be threatened by a new villain, Victor Vicious the Vitamin Stealer, who looked like a thinly reworked version of the TV series' Mervyn the Magician. Victor and his constantly mumbling evil minions would storm Granny's domicile in drastic attempts to snag Super Sugar Crisp for themselves. "Then I'LL have super power!" Victor would gloat. Hearing Granny's calls for help, Sugar Bear would eat fast and then develop a

When Sugar Crisp became the vitamin-fortified Super Sugar Crisp, Sugar Bear began turning to it in emergencies to gain "super power" to rescue Granny Goodwitch from the baddies.

larger chest and biceps as he declared, "Now I'm the bear of the hour!" A few energy-packed punches would dispatch Victor and his goons into the hazy yon, and Sugar Bear would help himself to Granny's cereal cupboard as in the old days.

The other Super Sugar Crisp series pitted Sugar Bear against a mob boss known as the Blob (no relation to the 1950s horror movie of the same title). The Blob spoke like James Cagney and also had his own group of thuggish assistants—but when Sugar Bear would make his vitamin-fortified appearance to foil their latest plot, the helpers would give up. "Get the bear!" the Blob would order; "You want him so bad, YOU get him," his cowardly crew would respond as they headed for the hills. In one commercial, obviously disgusted with his usual staff, the Blob infiltrates a mad scientist's laboratory and takes control of all the monsters. Upon Sugar Bear's arrival, the Blob orders them to "scare the bear." "With a face like that, YOU scare him!" the fiends shoot back. Every commercial ended with Sugar Bear admonishing the viewers to "get Super Sugar Crisp for super power," a claim that did not endear these spots to parental groups such as Action for Children's Television (ACT), formed in 1968. These "mothers from Boston," as network TV executives usually tried to dismiss them, would soon be having a greater effect on what their kids could see over the next few years.

But first, Post still had more new products up its sleeve—some of which succeeded beyond anyone's wildest hopes and others that crashed and burned without leaving a trace. In the latter category was Corn Crackos, twisty corn sticks introduced in 1968 with a mascot who looked suspiciously like the role might have been intended for Billy Bird. The resemblance ended with physical design; Cracko the Wake-Up Bird wore a Beatle haircut and had a Liverpool accent to match. He introduced Corn Crackos to sleepy non–morning people nationwide, with gags reminiscent of the Ruffles potato chip routines. The crunch from a mouthful of Corn Crackos was enough to cause the earth to tremble and produce panic in the streets. Corn Crackos lasted only a short while in the marketplace, probably because it tasted so much like all the other corn-based cereals out there, but Bob Traverse, the artist who designed the box, had another theory. "You've got a bird sitting on a spoon and it looked like a bowl full of worms. Who wants to eat worms for breakfast?"

Corn Crackos was soon forgotten, because at just about the same time that *Linus the Lionhearted* was ending its network run, Post introduced another new cereal that would go on to become the company's grocery store mainstay right up into the twenty-first century. No one saw such

Corn Crackos was not one of Post's more successful cereals. The mascot, Cracko the Wake-Up Bird, wore a Beatle haircut and spoke with a Liverpool accent.

longevity coming, of course, as the new product was initially the result of two seemingly unrelated events.

First, Post was getting ready to phase out Sugar Rice Krinkles—So-Hi had already been sent back home on a slow boat to China, and the company was keen on giving the sweet rice flakes a new tutti-frutti flavor. At the same time this was happening, over at Kellogg's, the eleven-year-old licensing agreement with Hanna-Barbera was coming to an end. Kellogg's had stubbornly clung to the studio's productions during the 1960s, culminating in a huge push for *The Banana Splits*, which starred costumed characters rather than animated ones. A few promotions connected with the *Dastardly and Muttley* cartoons would extend into 1970, but for all practical purposes, the Hanna-Barbera cast now consisted of free agents.

Somehow or other, Post decided to pick up the option on the characters from *The Flintstones*, which had been around since 1960 but oddly had never before been used to sell cereal. (During the program's original

After a profitable association lasting a decade, Hanna-Barbera's relationship with Kellogg's came to an end with the many promotions for *The Banana Splits* in 1968–69.

prime-time run, one of the main sponsors had been Welch's, which packaged its grape jelly in drinking glasses with Flintstones scenes on them.) The fruit-flavored Krinkles were renamed Pebbles, and the front of the fluorescent pink box featured an exuberant Fred Flintstone yelling, "Yabba dabba doo! It's the Flintstone cereal for you!" (Aside from Wheaties and its long-established tradition of having sports figures on the box, this might have been the first instance of cereal packaging *not* showing a picture of what was contained therein.)

Pebbles was so successful that shortly after its introduction, a chocolate-flavored version was added and the two brands were advertised side by side as Fruity Pebbles and Cocoa Pebbles. The TV commercials proved once and for all that no other company could match Post pebble for pebble in the "cereal thievery" theme. Fred and his friend Barney Rubble sparred endlessly over Barney's (usually successful) attempts to trick Fred out of his breakfast bowl. In one of the earliest examples, Barney shows Fred his new invention, a playground ride he calls the "more." While Fred spins uncontrollably on it, he quite logically, and loudly, asks why it is called a "more." "Because the more you ride it, the more I can eat your Pebbles!" chuckles Barney, prompting yet another chase across the prehistoric countryside.

As they had done from the original program's inception, Alan Reed and Mel Blanc supplied the voices of Fred and Barney. Most likely because of the expense involved, the other characters from the series appeared infrequently after the first few years. Both flavors of Pebbles were among the busiest premium suppliers in the grocery store, although most of the giveaways came in only three shapes: Fred, Barney, and pet Dino (rarely Wilma or one of the others would be included).

As mentioned above in regard to Super Sugar Crisp, Pebbles being tied so closely with *The Flintstones*, which was still appearing in reruns, would be one of the products that drew the ire of Action for Children's Television. Before we open up that old wound, though, in our next chapter we must backtrack and take a peek at what some of the smaller cereal companies were doing while General Mills was going cuckoo, Kellogg's was feeling grrr-eat, and Post was proving crime did pay—at least if it involved stealing cereal.

Stays Crunchy, Even in Milk

Quaker, Nabisco, and the Rest

Although Kellogg's, Post, and General Mills (in that order) were the three big gorillas in the cereal industry, from time to time other contenders would arrive to monkey around the neighborhood. With only an occasional exception, they all ended up being rather small bananas.

You may recall from many pages ago that the whole presweetened-cereal concept got started with the regional Ranger Joe product that became the "inspiration" (to put it kindly) for Post's Sugar Crisp. Well, in 1951, what was then known as Ranger Joe Wheat Honnies and Rice Honnies were still on the market and still using the staunch and steady Ranger on their plastic bags. This minion of the law was about to find himself forced into an early pension by a corporate merger.

The giant Nabisco company had not evidenced any previous interest in breakfast cereal other than its signature item, Shredded Wheat. In 1955, the temptation of the "already sugar'd" market (as it was spelled on the packaging) prompted Nabisco to swing a deal and purchase the whole Ranger Joe enterprise. Nabisco corrected Ranger Joe's spelling to Rice and Wheat Honeys, but evicted ol' Joe in favor of a more 1950s cartoon figure, Buffalo Bee. (Ranger Joe's name would continue to appear on the boxes for one more year before being allowed to fade into obscurity.)

Buffalo Bee became the representative for Nabisco's Rice and Wheat Honeys in 1955. As would occasionally happen with other mascot characters in the future, Buffalo Bee was merchandised outside the true cereal realm, notably with this comic book from Dell.

Buffalo Bee was one of those characters who knew he was a product spokesman and accordingly would talk back to the announcer or directly to the viewers at home. His "howdy, podner" accent was supplied by one of cartoonland's most famous voice actors, Mae Questel, who was simultaneously employed as the voice of Olive Oyl in the Popeye series. Questel was at her best when Buffalo Bee would foul up during his pitch, getting entangled in the rope he used to outline the shape of the cereal box. "Now I'll rope some words," he promises, cueing a flock of pigeons to swoop down. "I said WORDS, not BIRDS!" roars Questel. In another misadventure, Buffalo Bee demonstrates how he loads his guns with honey from the sweetest flower blossoms. When one of them refuses to produce any honey, he belatedly realizes it is only a plastic flower atop an irate old lady's tacky hat. Buffalo Bee buzzed TV spots for approximately ten years, but in 1965 was revamped into a much blander persona, Buddy Bee, who had neither the personality nor sticking power of the mite-sized cowpoke.

Despite the purchase and subsequent rehashing of Ranger Joe, Nabisco's primary cereal product remained Shredded Wheat. The company came up with a kid-friendly version known as Shredded Wheat Juniors, and in 1958 took full advantage of the current Space Age craze to give its product a trio of alien mascots. These were known as the Spoonmen: Munchy, Crunchy, and Spoonsize. Like Kellogg's more famous elfin trio, there was little to distinguish one Spoonman from another. Their most prominent physical feature was one most kids probably did not even notice: the Spoonmen's oval heads and projecting antennae atop their domes were a spot-on rendition of the famous Nabisco logo appearing on every one of the company's products.

The Spoonmen were the subject of many giveaway premiums, the most famous of which were the "spoon sitters." Many other cereals in the future would issue some variation on this concept: plastic figures that could perch on a spoon or hang onto the edge of a cereal bowl. Such a novelty was not enough to increase sales substantially, and the Spoonmen were sent back into orbit around 1960.

The lure of outer space was also a key component in another of the 1950s cereal companies, Ralston-Purina. Back in our first chapter, we saw how Ralston had been a powerhouse in radio advertising with its fictionalized adventures of the equally fictionalized Tom Mix. When it came time for television, Ralston jettisoned the wide open spaces for the final frontier. Ralston's *Space Patrol* series debuted in September 1950, less than a month before Kellogg's *Tom Corbett, Space Cadet*, making one

The intergalactic Spoonmen flew circles around Nabisco's Shredded Wheat Juniors in the late 1950s and early 1960s. Although they appeared as animated characters in the commercials, they were primarily designed to be plastic figures that perched on the handle of a spoon.

wonder whether all the spies and intrigue were actually confined to the TV studios.

While the hot-cereal version of Ralston had been the mainstay of the Tom Mix radio show, *Space Patrol* hawked the company's cold-cereal brand, Chex (formerly known as Shredded Ralston). Appearing in both wheat and rice flavors—Corn Chex would come along later—Chex took its name from the red-and-white checkerboard pattern that had long been Ralston-Purina's trademark for both its human and animal food. While *Space Patrol*'s Commander Buzz Corry and Cadet Happy were not Wheat and Rice Chex's advertising mascots in the same way that the Hanna-Barbera characters were conscripted into Kellogg's service, their colorful depictions on the Chex packaging made them seem much like true logos. *Space Patrol* premiums dominated Chex boxes until Ralston dropped its sponsorship in 1954.

The 1950s were an age of psychiatrists and psychoanalysis, so perhaps it was not surprising that Ralston next chose to advertise Chex with what can only be viewed as an extreme case of reverse psychology. Professor Checkerboard was a live-action pedagogue whose stovepipe hat and vest mirrored the red-and-white pattern, but that was practically the only thing cheerful about him. As Scott Bruce phrased it, the pedantic professor "gave kids a frightening glimpse of higher education." Although Checkerboard alternately promoted both the original hot Ralston and Chex, his pitches for the latter are what seem the oddest today. Rather than overtly entreating kids to eat Chex, Professor Checkerboard made the cereal into a sort of forbidden fruit:

> How do you dooooo, children . . . This is Professor Checkerboard, and I am glad to see you. But, children, what's this I hear? Have you been carrying tales to your mothers about Wheat Chex? Now, I've implored you NOT to eat Wheat Chex, because they were made on purpose just for grownups, and that when you were grown up you might enjoy the wonderful, yummy flavor of Wheat Chex. But children, I never, never said there was anything in Wheat Chex that was injurious to your digestive systems! No sirree! Scout's honor, children; everything in Wheat Chex is good, good, good . . . provided your mothers and fathers let you have some of THEIR Wheat Chex. You see, children, Wheat Chex are so good for you, so grand to eat, that the Ralston people have decided to reserve them for your mothers and fathers. Now, you can ask your mom and dad, but I, Professor Checkerboard, can make no promises. Be patient,

children; someday you will be grown up and then you can have all the Wheat Chex you can eat!

Perhaps realizing they had gone too far with this unappealing personality, Ralston did what few other cereal companies have tried: they gave Professor Checkerboard his own curtain call, and in one final spot he announced to the world that no one would ever see him again: "Children, I want you to remember me. Eat your Hot Ralston for my sake, and in memory of me, Professor Checkerboard. I shall be far away, children, but remember that I can see you, and I will always be watching to see that you eat your Hot Ralston."

(The specter of Professor Checkerboard was perhaps still hovering about a few years later, when another hot cereal called Maltex briefly advertised with a marionette called Professor Nutty. Rather than a stern taskmaster, Professor Nutty was a caricature of comedian Ed Wynn, with a goofy voice sounding much like another of Dallas McKennon's many characterizations. Professor Nutty made scant impression, if any, on the advertising world, even though he coined a new slogan for Maltex: "Nutty but nice.")

Professor Nutty, a somewhat creepy-looking marionette with some remote-controlled parts, briefly advertised the hot cereal Maltex.

Meanwhile, back at Ralston-Purina's headquarters at Checkerboard Square, St. Louis, the idea of marketing cold cereals to the children's market was allowed to cool for another decade. By the mid-1960s, even Ralston could not resist the siren call of the TV success of Kellogg's, Post, and General Mills, so Chex got a new spokes-character in the form of . . . and this had to be a "eureka!" moment for some ad executive . . . the Checkerboard Squarecrow.

Dancer Bobby Van was cast as the new character, which looked about as much like the *Wizard of Oz* scarecrow as the one who had advertised for General Mills' Country Corn Flakes earlier in the decade. (By sheer coincidence, when Walt Disney produced a short television segment based on the Oz books back in 1957, Van had supplied the voice of the Scarecrow.) The Checkerboard Squarecrow was a very engaging character and inspired some premiums that are highly sought after today, but his time in Chex's cornfield was somewhat limited. He was gone by the end of the decade, and Bobby Van died in 1980.

Ralston had not yet tried to horn in on the presweetened cereal market, and its first tentative step to do so occurred the same year the Squarecrow first climbed down from his pole. In 1966, Ralston added a flavored coating to Chex and called the resulting treat Mr. Waffles. This little-remembered product came in regular and banana flavors, and the even less-remembered logo character was named—what else?—Mr. Waffles. This gent was a jolly Englishman somewhat resembling TV actor Bernard Fox, with a bowler hat and a scarf bearing the red-and-white checkerboard pattern. Unfortunately for Ralston, Mr. Waffles never really managed to get off its beginning square, and the company seemingly swore off presweetened cereals until the 1970s, when its most famous brands would finally hit store shelves.

In the late 1950s, the same company that churned out the hot cereal Maltex came up with what it termed a "maple-flavored oatmeal," and enlisted famed animator John Hubley (one of the key creators of Mister Magoo) to produce the commercials. Hubley's contribution to cereal history was Marky Maypo, a Dennis the Menace–type brat who refused to eat Maypo until he saw someone else—normally his Mr. Wilson–like Uncle Ralphie—wolfing it down instead. Marky would then loudly bellow his catchphrase, "I WANT MY MAYPO!"

After the initial Marky/Uncle Ralphie confrontation, additional episodes introduced Marky's little sister, who managed to destroy their relative's carefully prepared dinner party as she searched for Maypo. TV commercial historian Jim Hall wrote, "Marky's cousin appeared in a new

Ralston's Checkerboard Squarecrow was played by Bobby Van, famed for his eccentric dancing. Both this character and the scarecrow who advertised General Mills' Country Corn Flakes earlier in the 1960s were obviously based on Ray Bolger's portrayal in *The Wizard of Oz*.

Marky Maypo badgered his relatives and friends for a taste of the "maple-flavored oatmeal" in John Hubley's highly stylized animated commercials of the early 1960s.

commercial in 1962. She was an emaciated beatnik who slinked around on limbs of rubber. Marky stared at her as if she had just dropped in from another planet." Marky Maypo premiums command premium prices on the collectors' market today, but the commercials are primarily remembered for his single "I WANT MY MAYPO" line, which delighted some viewers and grated on others like fingernails on a chalkboard.

Another hot cereal, Malt-O-Meal, took the evergreen Popeye/Cheerios Kid approach. Tiny tot Freddie, who wore a baseball cap over his eyes somewhat like Beetle Bailey, constantly outperformed his smug older brother by consuming Malt-O-Meal and then, for example, grabbing his sibling's baseball bat and smacking a home run, causing the bat to splinter. Although the theme was a tried-and-true one—and, of course, would be seen again—Malt-O-Meal just never captured enough public attention to make its commercials more than an occasional nostalgic footnote in advertising history.

Without any question, the company emerging as the biggest rival to Kellogg's, Post, and General Mills during the 1960s was the one that had started the whole cereal industry a century earlier. Quaker Oats was still doing well with its original oatmeal, of course, but had also expanded into the cold-cereal market with its Puffed Wheat and Puffed Rice. (These products inspired one of satirist Stan Freberg's most wickedly funny parodies with his bogus commercial for Puffed Grass: "You can always tell the Puffed Grass eaters in any crowd. They've got a green mouth.")

When the 1950s arrived, Quaker was still hanging onto the residue from its radio sponsorship days. Most of the decade was eaten up with promotions for the radio (and then television) program *Challenge of the Yukon*, better known by the name of its main character, Sergeant Preston. The Canadian Mountie and his dog, Yukon King, foiled one criminal plot after another, while also finding time to appear on the box fronts and on a seemingly endless series of premiums.

Much has been written about one of Quaker's most unusual Sergeant Preston promotions, but it deserves at least a brief mention here. In 1955, the company took out a lease on a piece of property in Canada's Yukon Territory and divided this barren wasteland into one-inch squares. Deeds to these peewee properties were included in boxes of Puffed Wheat and Puffed Rice, making each recipient a genuine Yukon landowner. At the time, no one thought of it as more than the usual advertising gimmick it was, but the interesting part came years later when people began coming forward with accumulated hundreds of the "one square inch of Yukon land" deeds, claiming they now owned enough property to move in, build a house, and take possession. Quaker was forced to backtrack and explain that nowhere was there any proof the one-inch squares were contiguous—and besides, the land had only been leased by Quaker for the duration of the promotion, so all of those would-be Sergeant Preston neighbors had to remain in their more mundane hometowns.

Another TV personality who had his grizzled face plastered across Puffed Wheat and Puffed Rice boxes was that cantankerous old coot of western movie fame, Gabby Hayes ("Yer durn tootin'!"). Quaker sponsored two different series known as *The Gabby Hayes Show*, in which the toothless old desert rat would tell tall tales, deliver American history lessons, and introduce chopped-up versions of old cowboy screen epics. Quaker offered enough Gabby premiums to keep the Pony Express busy hauling mail for years, but the product itself was still mired in the past.

Puffed Wheat and Puffed Rice were not presweetened cereals, and their TV commercials still relied heavily on the traditional image of their nuggets being "shot from guns" to the pounding tune of the "1812 Overture."

Finally, in 1957 Quaker got into the presweetened business with Sugar Puffs, which merely added a sugar coating to the existing Puffed Wheat—thereby creating basically the same product as Sugar Crisp, Sugar Smacks, and Wheat Honeys. Initially, reliable old Sergeant Preston and Yukon King were on hand to plug the new treat, but soon Quaker introduced its first two mascot characters specifically created for cereal advertising purposes: Mort and Wally.

What's that, you say? Who were Mort and Wally? It is true they (and their look-alike, taste-alike cereal) failed to make much of an impression. Mort was a moose and Wally was a walrus—despite the irony that another Wally Walrus was concurrently appearing in the Woody Woodpecker cartoons—who appeared on the box fronts and in newspaper and magazine ads. Their impact on Quaker's share of the presweetened cereal market was next to infinitesimal, but if nothing else, they proved to the company's marketing department that it took more than slapping cartoon characters onto a copycat product to make any real impression. The next time Quaker decided to introduce a new presweetened product, it would be the result of long weeks and months of careful study and analysis.

As the company's official history described it a few years later, "Market researchers determined the under-tens much preferred a crunch food. At the same time, the advertising department was developing a trademark strong in juvenile appeal: a salty, bumbling, comic old sea captain, an animated cartoon character surrounded by a fascinating coterie of friends, both human and piscatorial." For the first time, a cereal and its mascot character would share the same name, for the ultimate in brand identification. Quaker decreed its new product would be known as Captain Crunch.

Then came the job of finding an animation studio to produce this ancient mariner's exploits. Proposals were solicited from all the majors (including Disney, which had had quite enough of the TV commercial business and declined to even participate in the competition). The ultimate winner was Jay Ward Productions, even though the studio was already busily engaged in producing commercials for rival General Mills, not to mention its usual ration of cartoon series. Ward immediately set about developing a world for this new character, Captain Crunch, and rounded up his usual cast of voice actors. The mustachioed old captain would be played by Daws Butler—as much a double agent as Ward, since he was working so steadily in the Kellogg's commercials of the era. Writer Bill Scott, the voice of Bullwinkle and Dudley Do-Right, would fill in wherever an extra voice was needed, and the female member of the Crunch crew would be played by June Foray in her traditional "little girl" characterization. A flub by Foray gave the whole cereal campaign a new name: during one of the first recording sessions, she inadvertently mispronounced "Captain" as "Cap'n." Everyone liked the sound of it so much that the cereal and the character became Cap'n Crunch.

Speaking of voices and the people who spoke, TV viewers of the early 1960s might not have realized Butler's distinctive Cap'n Crunch delivery was actually an impersonation of a long-forgotten motion picture and radio character actor of the 1930s and 1940s: Charles Butterworth. Yes, his name sounds like he should have been enlisted to sell pancakes instead of cereal, but Butler loved to imitate Butterworth's slurring delivery that elongated the end of every sentence. He had used a variation of the voice in several earlier Jay Ward productions, but once it became attached to Cap'n Crunch, no one who heard it could even imagine it coming from another character's mouth.

The first commercials appeared in six test-market cities during September 1963. They introduced the world—or at least the world of those six communities—to doddering old Cap'n Crunch in his Napoleonic hat and droopy mustache. Also on board the good ship *Guppy* was the Cap'n's kid crew, consisting of (alphabetically) Alfie, Brunhilde, Carlisle, and Dave. There was also Seadog, a shaggy bipedal canine who was designed and animated a whole lot like Bullwinkle. Demonstrating his self-awareness as an advertising mascot, Cap'n Crunch constantly reminded viewers that the cereal "has to be good, because they named it after me!" He also hammered home the selling point the research brain trust had come up with: "It has corn for crunch and oats for punch, and stays crunchy even in milk." The test markets' responses were overwhelmingly positive, and the Cap'n was being nationally distributed by the end of the year.

Jay Ward and his merry madmen obviously had a great time concocting the Cap'n Crunch scripts, even though they were limited to sixty seconds to tell their stories. The commercials frequently depicted the Cap'n and the kids (say, wasn't that an old newspaper comic strip?) pulling ashore on some tropical island where they would introduce the cereal to the natives. In one such scenario, they find Robinson Crusoe living a life of luxury with a monkey-powered air-conditioning system and other such Flintstone-like innovations. When he claims there is nothing the outside world has that he could possibly want, the gang presents him

When Quaker Oats introduced Cap'n Crunch in September 1963, its success was astonishing. Besides being truly delicious, the product benefited greatly from the witty commercials from the studio of *The Bullwinkle Show*, Jay Ward Productions.

with a bowl of Cap'n Crunch. Crusoe is amazed that it stays crunchy even in coconut milk, but decides that now that he has a box of Cap'n Crunch, there really is no reason for him to return to civilization.

Even a friendly old salt like Cap'n Crunch had his enemies, though. Through commercial after commercial, battle-axe Magnolia Bulkhead (played by June Foray in her Marjorie Main/Ma Kettle voice) schemed to get Cap'n Crunch's cereal supply and also marry the old coot. The longest-running villain of the series was Jean LaFoote, "the barefoot pirate," whom Bill Scott played with a zany French accent. He, too, was determined to hijack the world's supply of Cap'n Crunch cereal, and frequently took the eponymous Cap'n hostage to accomplish his dastardly goals. In their first encounter, LaFoote holds Cap'n Crunch at swordpoint to force him to tell his secret:

LAFOOTE: But how does eet stay crunchy, even after you pour on ze milk?
CAP'N: Oh, that's a secret.
LAFOOTE: (yelling) I KNOW eet's a secret! And I must have eet! Give eet to me!
CAP'N: Very well . . . Seadog, give it to him.
(Seadog cuts the rope on the ship's anchor, which squashes LaFoote into the ground)
(Meanwhile, Cap'n Crunch enters a nearby telephone booth)
CAP'N: (on phone) Hello? Quaker Oats Company? Say, how DOES my cereal keep its crunch? Uh huh . . . yes . . . I thought so.
LAFOOTE: Well?!
CAP'N: It's just as I thought . . . it's a secret.
(LaFoote begins throwing a tantrum)
CAP'N: Let's go, Seadog . . . I hate to see a grown pirate cry.

Ward's cartoon series were well known for their satire on all forms of show business, so it must have made him and Bill Scott positively giddy when they cooked up one particular Cap'n Crunch spot. It spoofed the omnipresent Timex watch commercials in which announcer John Cameron Swayze would submit the timepieces to various torture tests in order to prove they could "take a licking and keep on ticking." In Ward's parody, a Swayze caricature announces they have strapped Cap'n Crunch to the rudder of his ship and are going to dunk him into Lake Michigan, which has been drained and filled with milk, to find out if he really will stay crunchy:

SWAYZE: Cap'n Crunch, do you have anything you would like to say to our home audience?

CAP'N: As a matter of fact, yes.

SWAYZE: And what is that?

CAP'N: HELP!

After the Cap'n is slapped around by the rudder a few times, the Swayze clone tries the cereal and finds, yes, it has stayed crunchy. One could probably see the typical Jay Ward punch line coming, though:

SWAYZE: And how did you come through it, Cap'n?

CAP'N: Oh, fine, I guess . . . I'm gonna have to get a milk-proof watch, though . . .

Ward's influence was felt in some of the early Cap'n Crunch premiums, too. There was a series of giveaway comic books with titles strongly reminiscent of the closing gags in the *Rocky and His Friends* episodes, such as "I'm Dreaming of a Wide Isthmus," "Cap'n Crunch and the Picture Pirates, or, the Rogue's Gallery," and "Seadog Meets the Witch Doctor, or, I'll Bewitch You in a Minute." Historian Keith Scott reports that at some point there was talk of producing a half-hour Cap'n Crunch series for Saturday mornings, but for one reason or another, the idea fell through. (Perhaps it was because Post beat them to it with *Linus the Lionhearted*.) Nevertheless, the commercials managed to introduce a Cap'n Crunch theme song, to the tune of Gilbert and Sullivan's "When I Was a Lad":

Sing a song of Cap'n Crunch,
The sugar-sweet cereal that's fun to munch,
It's great for breakfast or even lunch,
And no amount of milk will make it lose its crunch.
So all ashore that's going ashore,
Get Cap'n Crunch at your grocery store.

Now that Quaker understood the highs and lows of the children's cereal market, the company wasted no time in getting another product onto the market—or actually, two products. Determined to use the uniqueness of their initial letter as much as possible, Quaker developed a pair of cereals called Quisp and Quake and once again depended on Jay Ward's nutty brand of genius to help.

As with Cap'n Crunch, the names of the cereals would be the same as those of their representative characters. The big innovation this time around was the concept of Quisp and Quake being competitors, with each out to sabotage the other's market share. The two mascots were about as different in every way as Ward's creative team could make them. Quisp, a resident of Planet Q (there's that letter again), was a pink alien with a propeller atop his bald dome. His eyes constantly rolled around their sockets as if he were a few craters short of a planet, and his voice (supplied by Daws Butler) was so much like comedian Jerry Lewis that it is impossible to effectively capture its delivery in print. Quake, on the other hand, was a burly, muscle-bound strongman from the center of the earth, whose powerful fists could move tons of rock with barely an effort. His deep bass voice was that of William Conrad, the veteran dramatic actor who had proven his knack for comedy as the frantic narrator on *Rocky and His Friends*.

The rivalry began in the very first commercial, which was hosted by an announcer played by Paul Frees:

ANNOUNCER: Ladies and gentlemen, before your very eyes the Quaker Oats Company will now introduce two new cereals.

QUISP: I'm Quisp, the quisp new cereal from outer space! The biggest-selling cereal from Saturn to Alpha Centari! Quisp is sugary sweet and vitamin charged to give you QUAZY energy! (A box of his cereal hits Quake on the head) What's new with you?

QUAKE: I am Quake, the POWER cereal from inner space! (He tunnels to the center of the earth) Here at the earth's core I make Quake, with deep-down sweetness and vitamins to give you the power of an earthquake! (He leaps back to the stage high above) Get QUAKE!

QUISP: Quisp is better!

ANNOUNCER: Fellows, why not leave it to the kids out there? Take sides with either Quake or Quisp . . .

QUAKE: Or Quake.

QUISP: Or Quisp.

ANNOUNCER: Two new cereals from Quaker . . . sort of a breakfast feud.

After the opening salvo, the commercials fell into their regular pattern. An adventure would befall either Quisp (in space) or Quake (on or under terra firma), and the hero would have to deal with it in his own fashion. The other would turn up at the end to heckle the star of that particular spot. Quisp, being in a more solitary setting, usually ended up performing alone, but Quake soon introduced his doting "mama," Mother Lode,

Jay Ward's commercials for Quisp and Quake were groundbreaking in their own way. Never before had two mascot characters (and their namesake cereals) from the same company been depicted as enemies, with each out to convince the public that his rival's product was not worth eating.

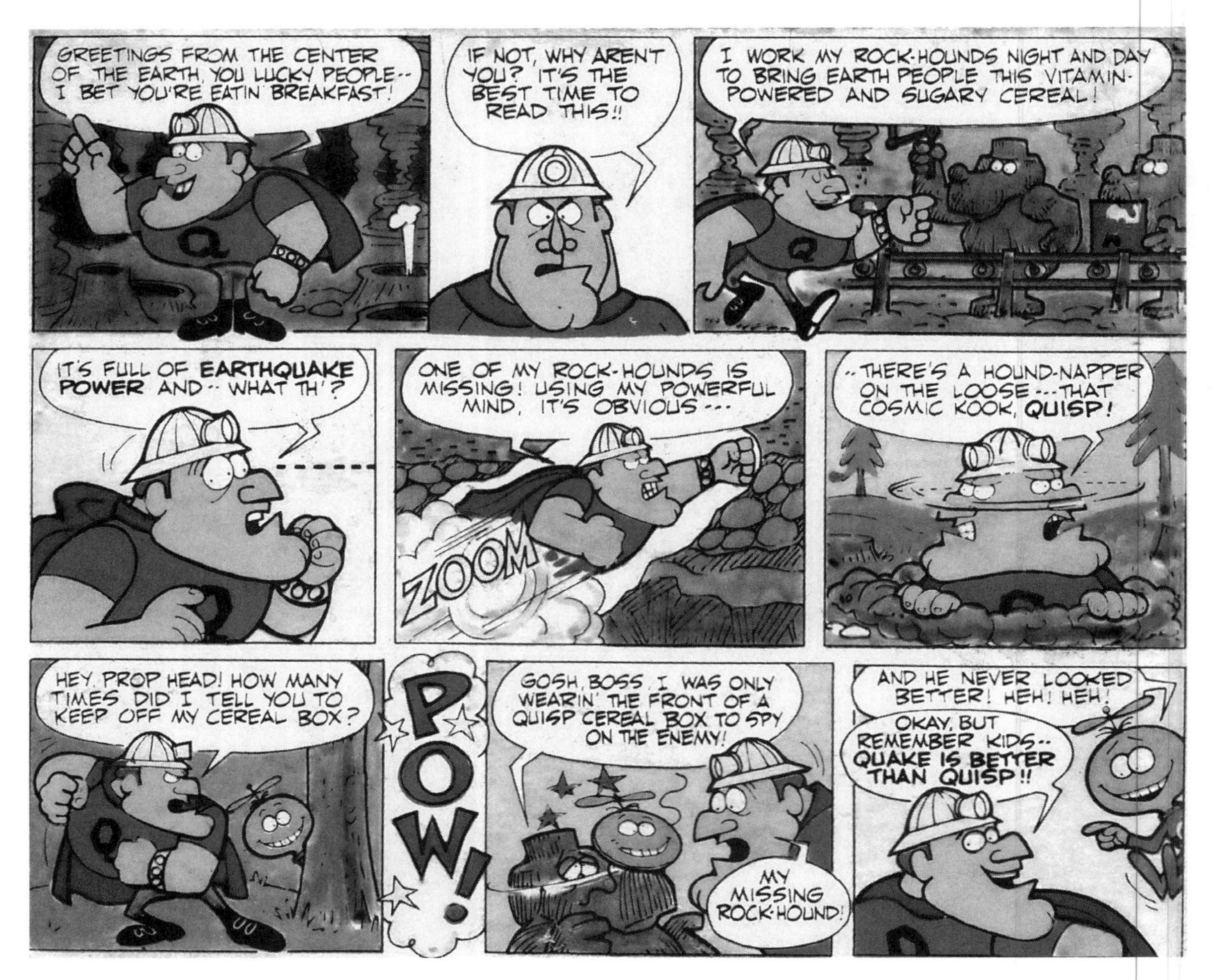

This comic strip that appeared on the back of a box of Quake not only demonstrates Jay Ward's trademark self-referential humor but also depicts the type of brute force that eventually prompted Quaker to soften Quake's burly image.

who, like all mothers, still thought of him as her little boy. On occasion, Quisp and Quake would team up to defeat a common foe, but they would always be back to arguing before the sixty seconds were over.

This rivalry was entertaining, but its choice of opponents had an unexpected drawback. Although Quake was obviously supposed to be a good guy, even wearing a cape like a typical superhero, his burly build made him seem more like a Bluto-type bully when he was pitted directly against the minuscule Quisp. Market research among kids found they were actually frightened of what the big guy might do if Quisp ever pushed the wrong buttons for him, so toward the end of the 1960s, steps were taken to soften his image.

One commercial started out much like all the others, with an airplane in distress headed straight for a craggy mountain, Whyntchatakea Peak (a Jay Ward name if there ever was one). Down below, Quake springs into action, although both Quisp and Mother Lode try to warn him he will never make it in time. Quake replies that he is using his "new and improver machine," leaping into its gears. "But I like you just as you are," Mother Lode protests, just before Quake pops out the other end in a new, less threatening form. Gone are his bulging biceps, massive chest, and miner's hat; now he has skinny arms and legs, a freckled, boyish face, and a cowboy hat. "How's THIS, Mama?" he gloats. "Like I said, I like you just as you ARE!" she admiringly replies. Of course, the "new, improved" Quake is able to break the top off the mountain and allow the plane to fly by undamaged, and that was the last anyone saw of his original design.

The Quisp/Quake competition continued, but now the latter combatant was not nearly so threatening. Quaker Oats was on a roll, and its next product was sent into test market mode in 1968. This was Frosted Oat Flakes (obviously a Quaker version of everyone else's frosted corn flakes), and the cartoon mascot was a young boy named Quincy Quaker. He was played by Dick Beals, fresh off his work on the Kellogg's commercials with Corny the Rooster, but even the presence of his always-delightful delivery did not help. Unlike Cap'n Crunch, Frosted Oat Flakes never got beyond the test markets, and Quincy Quaker became a casualty.

Quaker decided to stick to what it knew best, and in 1969, the first of a long line of Cap'n Crunch spin-off cereals was introduced. Its "discovery" was documented in a commercial in which the *Guppy* runs aground on a mysterious uncharted island, and Cap'n Crunch and his kid entourage go exploring:

CAP'N: Now, careful, crew . . . goodness knows what we'll find on this island.

BRUNHILDE: Goodness is right, Cap'n . . . just taste these red berries!

CAP'N: Why . . . they're crunchy!

ALFIE: They taste like strawberries!

CAP'N: But they're crunchy! Crew, we've discovered Crunch Berry Island! Oh, just wait'll I tell the folks at Quaker!

And so was born Cap'n Crunch with Crunch Berries, in a blindingly yellow box to distinguish it from the red box of the original. Originally, the front of the box depicted the Cap'n brandishing his sword in front of a bush composed of pieces of both types of cereal, original Cap'n Crunch and the new strawberry-flavored Crunch Berries. Before long, Jay Ward had introduced another character into the commercials, the yellow-and-red-spotted Crunch Berry Beast, or C. B. for short. This creature of indeterminate species spoke only in "bloop, bloop, bloop" sounds but went as berserk over Crunch Berries as Sonny the Cuckoo did for Cocoa Puffs.

As we shall soon see, Crunch Berries were only the beginning. Throughout the 1970s, Cap'n Crunch would produce more ancillary flavors than any other cereal before or since. However, all was not fun and games for the cereal industry during the decade, and we are about to examine how a rising tide of criticism changed the way such products were marketed to impressionable tots—those were you and me, folks.

Part of a Balanced Breakfast

The 1970s Get Serious

We have already touched briefly on the late 1960s rise of such parental groups as Action for Children's Television (ACT). These groups found much to criticize when it came to children's programming, be it of network, syndicated, or local origin, but their biggest gripes boiled down to two huge topics. They felt kids' TV was fraught with violence, whether it was the science fiction of the "weirdo superhero" cartoons or (and this is what stuck in many people's craw) the fantasy bops, bangs, and booms of the Warner Bros. and Popeye cartoons. None of this had anything directly to do with the cereal companies whose commercials interrupted all of the above types of shows equally—but ACT's other big axe to grind came awfully close to the cereal manufacturers' necks.

One of ACT's stated goals was the total elimination of *all* advertising during children's viewing time. Their much-cited ideal was *Sesame Street*, the public television show debuting in November 1969. Network programmers agreed that while *Sesame Street* was a worthy experiment in educational television, it was not practical to expect the rest of the industry to follow its model. The specter of no sponsors to lend their financial backing to network kids' shows would be sure to spell doom for the genre should it take place, but at least some concessions were made to buy some time.

ACT liked to point out the dubious nutritional value of the presweetened cereals, which they claimed often contained more sugar than any other ingredient. ACT brought in learned experts to solemnly testify that children could get more nutrition out of eating the boxes in which the stuff was packaged. Kellogg's, General Mills, and Post reacted by adding a line that would become familiar in every cereal commercial hence: whereas previously the cartoon characters ate their represented product alone, now each cereal became "part of a balanced breakfast," implying it was more nutritious if shown with orange juice, toast, fresh fruit, and other side dishes.

There were also some opinions that, frankly, made ACT's scholarly allies look like they were somewhat less intelligent than the children watching at home. They objected to the many campaigns showing characters, either live or animated, obtaining super strength or other powers from eating sugared cereals. Now, most kids were savvy enough to realize the difference between reality and cartoon exaggeration for comedy effect, but the experts did not—so that long-established shtick dating back to the mid-1930s radio days was soon put out with yesterday's trash.

Another objection the parental groups had was to the hosts or primary characters of children's shows actually delivering the commercials, or at least introducing them. Under pressure, the National Association of Broadcasters (NAB) issued a code, adhered to by virtually every network and TV station in the country, stating that such close connection between sponsors and stars would be severely frowned upon. This would have spelled the end of *Linus the Lionhearted* had it still been in production, but the commercials most directly affected were the ones for Post's Fruity and Cocoa Pebbles. One expert hired to analyze Saturday-morning cartoons claimed she saw Fred and Barney get out of their car to walk out of the scene, only to immediately turn around and begin pitching Pebbles cereal. Again, this brain must not have had the attention span of the kids, as there would never have been a transition from program to commercial as seamless as she described, but the claims were taken to heart. It was impossible to separate Pebbles from the Flintstone characters, so the commercials were allowed to continue as long as they were not shown during episodes of *The Flintstones* or any of its several Saturday-morning sequels.

The networks responded to the oft-deserved, sometimes-overblown praise for *Sesame Street* by announcing that they would step up the number of their own educational offerings. Kellogg's proved it was a team player by joining forces with ABC-TV to bankroll *Curiosity Shop*, which

This anonymous chief served as the mascot for Sugar Smacks during that brand's transition period between the Smackin' Brothers and Dig'em the Frog. Notice the premium offer connected with Kellogg's new Saturday-morning series *Curiosity Shop*.

These Kellogg's character stencils were offered in 1970. Sugar Pops' Whippersnapper had not yet lost his bullwhip, but this was already an outdated design for Toucan Sam.

premiered in September 1971. Very much a *Sesame Street* reimagined for grade-school students rather than preschoolers, the show offered a mix of live actors, animated characters, and oversized puppets. It was produced by Warner Bros. cartoon veteran Chuck Jones, and featured the voices of such legends as Mel Blanc, June Foray, Les Tremayne, and Robert Holt. In the new Saturday-morning climate, it would have been unthinkable for the motley *Curiosity Shop* crew to pitch Kellogg's products directly, but the company certainly spared no effort in stuffing the cereal boxes full of premiums from the show.

By the time Kellogg's sponsored *Curiosity Shop*, several of those boxes looked different than they had when we last saw them in 1969. For example, Sugar Pops' Whippersnapper had been disarmed of his bullwhip, and was now a standard Dodge City cowboy. Sugar Smacks had kept the Smackin' Brothers, still with Robie Lester's voice, around as long as possible, but with the new howls to rein in cartoon violence, somehow the prospect of two kids whaling away at each other just did not seem to fit. Without warning, the brothers went to the Old Mascots' Home while still at a young age and were replaced by a nameless Native American chieftain whose only real purpose seemed to be to balance out Sugar Pops' cowboy.

Apple Jacks had toyed only briefly with the "apple car" logo that had replaced the bully-baiting, fruit-headed Apple Jack character. By 1971, the chartreuse boxes bore the images of a boy and a girl who appeared to have been drawn by a crayon-wielding first-grader. Never given official names, the Crayon Kids, as they were usually called, did not have any personality and played little to no role in Apple Jacks' TV advertising, but apparently Kellogg's was happy enough to keep them around in the box design well into the 1980s.

Cocoa Krispies were still being touted by bucktoothed Ogg the Caveman, but not for long. The process of replacing Ogg was going to be a long and twisting one, though. In late 1972, the Cocoa Krispies commercials began featuring a small, rotund, brown elephant who wore huge

The Crayon Kids were the latest, and longest-lasting, mascots for Apple Jacks. Their presence seemed coincidentally appropriate on this particular box, which promoted Kellogg's offer of a *Curiosity Shop* paper coloring roll.

This 1974 Kellogg's newspaper ad is another example of mascots in flux. Notice the illustrated Cocoa Krispies box depicting the new character Tusk Tusk the Elephant, while the decals being offered still feature Ogg the Caveman.

spectacles and spoke with what sounded like the voice of Wally Cox. To date, no evidence has turned up to prove conclusively that Cox really was the original voice of this elephant, but since Cox died unexpectedly of a heart attack in February 1973, his participation—if it existed at all—was short. The elephant was fond of clucking "Tusk, tusk" at everyone, for no good reason, so it became the new Cocoa Krispies mascot's name: Tusk Tusk the Elephant.

For the remainder of the 1970s, Tusk Tusk's voice was provided by Paul Winchell, but it seems the Neanderthal Ogg found it difficult to relinquish his place in the cereal evolutionary chain. In 1973, well after Tusk Tusk had been introduced, the Kellogg's lineup offered a set of plastic bicycle "license plates" with the various mascots' pictures, and it was Ogg,

not Tusk Tusk, who represented Cocoa Krispies in that series. Even as late as 1974, Ogg was enshrined in a set of giveaway decals of the Kellogg's cast, even though the same ads depicting the decals also featured the Cocoa Krispies box with Tusk Tusk looming large. Apparently by 1975 or thereabouts, Ogg had finally crept back into his cave, leaving the pompous pachyderm to trumpet his approval of Cocoa Krispies for the next several years.

Another character who appeared in the bicycle license plate series was Newton the Owl. Yes, we know what you're going to say: "WHOOOOO?" Oh, come on—you remember Newton and the cereal he represented, Cocoa Hoots, don't you? Anyone? Well, the best way to describe Cocoa Hoots is to say they were the same as Froot Loops except chocolate-flavored,

The little-remembered Newton the Owl, mascot for the even more obscure Kellogg's product Cocoa Hoots, made one of his few premium appearances in this set of plastic bicycle license plates.

which is probably why the Leo Burnett agency came up with another bird character to advertise them. In sharp contrast with Toucan Sam's upper-class British accent, Newton spoke with the voice of famed rural comedian Pat Buttram, whose hit sitcom *Green Acres* had just been canceled the previous year. Poor old Newton. After making it onto a pitifully small number of premiums, he did not achieve anywhere near the same stardom as his feathered cousin Sam, and he soon flew silently away into the darkness of obscurity.

The next Burnett creation for Kellogg's had a much longer career, which is a bit surprising considering how much he was a product of his own time. As we have been seeing all along, Sugar Smacks was perhaps the most fickle of all the Kellogg's cereals when it came to finding a niche for its advertising. Tony the Tiger, Toucan Sam, and Snap!, Crackle!, and Pop! held down their posts for decades, but Sugar Smacks had staggered from Ringling Bros. clowns to a circus seal to Quick Draw McGraw to two fighting brothers to a Native American chief who had no known name (and probably also rode through the desert on a horse with no name). The character who hopped into Sugar Smacks ads during 1972–73 would not have any more logical tie-in to the sugared wheat puffs than any of the others, but for one reason or another, his tenure has lasted longer than the careers of all his forerunners put together.

If you are up on your early 1970s lingo, you will know one of the "cool" phrases was "dig" for "like" or "appreciate," as in, "What do you think of that music?" "Oh, man, I really dig it." Using this and not much else as a basis, the Burnett crew thought up a frog who dressed like a hip 1972 schoolkid and bounced around croaking "dig 'em, dig 'em" on behalf of Sugar Smacks. Since there seemed to be no other logical name for him, Dig'em the Frog became his moniker.

Dig'em's deep bass croak came from the throat of Robert Holt, one of the veteran cartoon voice actors who had contributed to Kellogg's *Curiosity Shop*. (A few years later, Hanna-Barbera would ask Holt to supply the same half-burping voice for their character the Great Grape Ape, who constantly mumbled his own name, "Grape Ape, Grape Ape," in much the same fashion.) Dig'em, with his baseball cap, denim jacket, blue jeans, and sneakers, certainly had to have been one of the most contemporary cereal mascots ever introduced. Because his linguistic abilities were limited, Dig'em spent most of his on-screen time bouncing around and giving live-action kids high fives to the tune of the new Sugar Smacks jingle: "Gimme a smack and I'll smack ya back."

About the time Dig'em was being introduced, there was a rumored Kellogg's project that was never completed. Robie Lester, who had been providing voices for the company's commercials for the past ten years, related in her posthumously published autobiography how she had been working to develop a new mascot character whose name was Nutrina. What cereal Nutrina was to have been associated with was not part of Lester's memory, but she did recall that comedian Jimmy Durante was also involved in the Nutrina campaign. (Durante had already done a series of Corn Flakes spots in the late 1960s.) However, while Nutrina was still in development, Lester had to undergo two open-heart surgeries, which effectively ended her acting career. She said Durante frequently called the hospital to inquire about her prognosis, but by the time she was able to consider returning to work, Durante himself had fallen ill. Durante died in 1980 and Lester died in 2005, so the story of who or what Nutrina was will have to forever remain a "lost chapter" in Kellogg's history.

Sometimes a product can be a "lost chapter" even if it makes it into the grocery aisle and onto TV. During the 1970s, Kellogg's continued to do wonderfully with its established product line, but its track record was not so stellar when it came to introducing new cereals. Perhaps this was

Although Dig'em the Frog was firmly a product of his own time, with his sneakers, denim jacket, and hip vocabulary, the little green hopper has gone on to become the longest-lasting mascot Sugar Smacks ever had.

inevitable, as the combined efforts of all the cereal companies had pretty much ensured that all possible taste and flavor combinations had already been tried, and anything else could only be a rehash of what had been done before. That was no doubt the major drawback with Kellogg's Corny-Snaps, introduced in 1975. The presweetened corn/oats combination was identical to Cap'n Crunch, although the mascot character was certainly different.

Corny-Snaps was advertised by Snappy the Turtle, who dressed like Zorro and rode through the West carving an "S" (for Snappy) everywhere he delivered the new cereal. The most creative thing about Snappy was the agency's choice of a voice performer for such a heroic figure. Snappy spoke with the tones of famed radio, movie, and TV nebbish Arnold Stang, which made him sound like he was caught halfway between crying and yodeling. Stang was undeniably a talented actor and comedian, but Snappy was not the most memorable of his many roles, and the cereal he represented was discontinued after only a brief period.

Another product introduced in the mid-1970s was Frosted Rice, filling the gap left when Post's Frosted Rice Krinkles had morphed into Fruity

Snappy the Turtle, with the cracking voice of Arnold Stang, represented Kellogg's short-lived Cap'n Crunch clone known as Corny-Snaps.

Tony the Tiger's son, Tony Jr., briefly stepped out of his old man's shadow in the late 1970s and early 1980s to roar his approval for Frosted Rice. Here he can be seen with Kellogg's box lineup of circa 1980.

and Cocoa Pebbles. Since Frosted Rice was so similar in theme to Sugar Frosted Flakes, it was only natural that the new cereal was represented by Tony Jr., the little-boy tiger who had been around since the earliest Frosted Flakes boxes. He no longer had Hal Smith's impersonation of a kid voice, but was instead played by a real—though now unidentified—child actor. His commercials presented him as part of an otherwise live-action world, living in suburbia, but with Papa Tony nowhere in sight. Tony Jr. would interact with live kids, mimicking his old man by announcing (loudly) that Frosted Rice was "so good, it makes ya RRRRRRROAR." In the particular case of Frosted Rice, the product was more successful than the mascot. During the 1980s, Tony Jr. was sent back to his family unit, and Frosted Rice became just another flavor of Rice Krispies, domain of Snap!, Crackle!, and Pop!

(While we are on the subject of Tony's family, mention should be made of an often-broadcast commercial during 1974, which in the Chinese New Year calendar was the "Year of the Tiger." It was too great an opportunity for Kellogg's to let pass, and the elaborately animated spot depicted Tony returning to the jungle for the celebration with all his animal friends. Unnoticed by anyone except cereal historians, the opening line, "Look! Tony's come home!" was delivered by Katy the Kangaroo, whom Tony had left behind way back in 1953.)

Sugar Pops still had the Cowboy Formerly Known as the Whippersnapper on the box as late as 1977, but during that year, the cowpoke was replaced by . . . another cowpoke. (One thing is for certain: with its lineage from Wild Bill Hickok and Jingles, through Sugar Pops Pete, to the Whippersnapper, Sugar Pops deserved some sort of credit for sticking with the western theme in all its advertising.) At the same time that the new character, a 20-gallon-hat-wearing bronco buster named Big Yella, was introduced, the product name on the boxes officially became Sugar Corn Pops, its alternate logo since the days when Guy Madison and Andy Devine were appearing in the commercials. Big Yella's reign over Sugar Corn Pops lasted less than three years; at least one problem was that his voice and mannerisms were modeled after John Wayne, and after Wayne died of cancer in 1979, using a parody of him to sell cereal just did not seem fitting. Big Yella's replacement would get completely away from the western image, but we will come to that story—and character—later.

Kellogg's last, and perhaps least successful, new product during the 1970s was Crunchy Loggs. Now, if you remember Cocoa Hoots or Corny-Snaps, you might possibly remember Crunchy Loggs, but that is not guaranteed. Crunchy Loggs were largely a reincarnation of Corny-Snaps with strawberry flavoring added, probably giving them a taste much like Cap'n Crunch's Crunch Berries. The little pink logs were advertised by Bixby Beaver, who dressed like a lumberjack and spoke through gigantic buckteeth that might have frightened Bugs Bunny. Bixby's time was mercifully brief, with some sources claiming Crunchy Loggs lasted only a year.

For now, we must leave Kellogg's to its own devices, both successful and otherwise, and turn our attention to General Mills, which had outpaced Post to become the runner-up in animated mascots. The old favorites, Lucky the Leprechaun, Sonny the Cuckoo, and the Trix Rabbit, were still going strong, but in 1971 the company introduced two new

Big Yella, a caricature of John Wayne, might not have been tall in the saddle, and his days as the logo for the newly renamed Sugar Corn Pops were pretty short, too.

cereals whose names and ad campaigns were obviously inspired by what was happening over at Jay Ward's Quaker Oats camp.

General Mills had already pioneered the "marshmallow bits" formula with Lucky Charms (and their floparoo follow-up, Wackies), and these new products would build on this previous success by making the cereal and marshmallows either chocolate or strawberry flavored. For the first time, General Mills' products would share the names of their mascots, as had been done with Cap'n Crunch. It was time to unlock the forbidden door to the world of the undead and release Count Chocula and Franken Berry.

We have seen plenty of examples of characters who were timeless (Tony the Tiger, Toucan Sam, the Trix Rabbit, and many more) and others who attempted to be relevant to their time (Dig'em the Frog, the Nabisco Spoonmen, and so on). General Mills choosing a monster theme for its new campaign was also a reaction to pop culture, only it came a decade late.

The classic movie monsters of the 1930s had first reached TV screens in the late 1950s, when local stations in hundreds of markets began scheduling them as "Shock Theater" late-night offerings, usually with a ghoulish host who was a daytime kids' show emcee literally moonlighting. By the early 1960s, ten years before Count Chocula and Franken Berry went on the prowl, monster merchandise was scaring up huge profits with Aurora's successful line of model kits. There were monster-themed board games, Soaky toys, and even a line of jigsaw puzzles that had to be quickly withdrawn from the market because their artists got a little *too* carried away in painting their horror-scene images. By the middle of the decade, sitcoms such as *The Munsters* and *The Addams Family* had learned how to make the ghouls more funny than scary, and cartoons got in on the frightening fun with such series as Hanna-Barbera's *Frankenstein Jr.* When General Mills decided to go in that direction, they were not exactly robbing the grave, but the initial monster mayhem had died down at least somewhat by the early 1970s.

Since no one had ever seen characters quite like these used for cereal advertising before, the first year's commercials for the "Monster Cereals," as General Mills referred to them, were a little inconsistent. The running gag, lifted from the Quisp/Quake competition—which was beginning to wind down—was that Count Chocula and Franken Berry would constantly be at each other's throats (where else would the two creatures be?) as each tried to promote his cereal over the protests of the other. It did not take much imagination to figure out that Count Chocula's voice would be an imitation of Bela Lugosi, supplied by actor James Dukas. Likewise, Franken Berry would have to sound like Boris Karloff, and the ubiquitous Robert McFadden was brought in for the job.

The first two commercials put the main emphasis on one or the other of these Halloween haunts, with the remaining one making a cameo appearance. Each of these early spots was rooted in classic horror movies much more so than future installments, although they had a lighthearted mood about them. In his debut, Count Chocula was seen rising from a cardboard box in a grocery store cereal aisle, as if it were a coffin. "Don't be afraid," he soothed viewers who might have been about to reach for the TV dial. After the Count gave his pitch, Franken Berry popped in from nowhere just to interject that his strawberry-flavored cereal was better. Conversely, Franken Berry's first appearance depicted the pink behemoth on a slab in a mad scientist's laboratory, being brought to life with a jolt of lightning. "Ah, what a beautiful day," he cooed, bringing back memories of similar gags in the *Munsters/Addams* tradition.

Most of the future commercials would show the two monstrous mascots attempting a semi-peaceful coexistence, even as their jealousy toward each other burbled under the surface. When Franken Berry would attempt to have a picnic at midnight in the cemetery, Count Chocula would appear from behind a tombstone to tout his cereal's superiority. Unlike the similar escapades of Quisp and Quake, neither character ever came out as a clear "winner" in these battles of wits. Instead, each spot would end with their argument being interrupted by some mild outside source—a stray black cat, an owl, a human child just passing by—which would frighten the two cowardly creatures half to death (or, presuming they were already dead, half to life), abruptly causing the duel to end in a draw.

After a year of such contests, a new element was thrown into the monster mash-up in 1972. In one commercial, Count Chocula and Franken Berry are busily setting up a breakfast table in their (jointly shared?) gloomy old castle, when there is a knock at the door. They are initially delighted to assume "someone wants a balanced breakfast," following the new rules laid down for advertisers. Their visitor turns out to be a small blue ghost, and their mood turns from benevolent to malevolent when the new arrival announces his name as Boo Berry, and he is there to advertise his own new blueberry-flavored cereal with marshmallow bits. Apparently if there is one thing the Count and the monster can agree on, it is that they will put up with no further competition, and they summarily

In their first commercial, Count Chocula and Franken Berry argued the merits of their respective cereals in a typical supermarket. Future stories would place them firmly in a more familiar Transylvania-rooted world.

toss Boo Berry back into the haunted wilderness outside. The punch line, of course, is that Boo simply passes back through the closed door as if it were not there, sending the two big bullies screaming away.

Since the two preexisting monsters each had a celebrity-imitation voice, it stood to reason that Boo Berry would, too. His voice (Robert McFadden again) and physical appearance were modeled after famed screen creep Peter Lorre—and unlike Count Chocula and Franken Berry, who were all bluster, there was ample evidence that Boo Berry had *not* been one of the good guys while he was alive. In those earliest appearances, his ghostly form was laden down with chains, in true Jacob Marley fashion, indicating he was now reaping the eternal punishment he had sowed with his misdeeds. The big difference was that instead of Marley's metal moneyboxes, Boo Berry's chains were festooned with boxes of his cereal.

The addition of Boo Berry meant Count Chocula and Franken Berry more often than not had a common enemy to face, relieving the tedium of their constant one-upmanship. In one of these three-way contests, the two big lugs are at the beach during what appears to be a hurricane, arguing as usual, when Boo Berry shows up. Working fast, they wrap him in a

General Mills' third "Monster Cereal," Boo Berry, was introduced in 1972. The small blue ghost, who spoke like Peter Lorre, immediately made himself a pain to Count Chocula and Franken Berry, who schemed to eradicate the spook and his cereal.

convenient rubber toy and toss him into the stormy ocean—only to have him rise from the sand in front of them, frightening them into submission once again.

A fourth, and less legendary, member of this unholy group arrived on the scene in 1973. By this time, Count Chocula and Franken Berry are seen sleeping in adjacent beds, but they are awakened in the middle of the night—when, one would assume, they should normally be out on the prowl—by a distant howl sounding like "FRUUUUUUUIT." It belongs to a new character, Fruit Brute, who is undoubtedly intended to be a werewolf, although we never see him changing into human form or back again. Instead, he is merely there to introduce his own namesake cereal,

Fruit Brute, a jolly werewolf, joined the General Mills cast in 1973. He was probably the least successful of the ghoulish group, although he did at least get to be offered in the form of this rubber squeeze toy.

which features the marshmallow bits in assorted fruit flavors. Logically, one might assume Fruit Brute should have the voice of Lon Chaney Jr., but instead he was given a more generic "gruff" voice that might or might not have belonged to Robert McFadden.

It would appear that General Mills was convinced there was no such thing as having too much of anything, so they introduced yet another "cereal rivals" story line involving two non-monster characters, Baron Von Redberry and Sir Grapefellow. This short-lived series rode the coattails of a World War I craze largely ignited by Snoopy's comic-strip (and hit-song) battles with the Red Baron and kept alive by the movie *Those Magnificent Men in Their Flying Machines*. In case it was not obvious from his name, Baron Von Redberry was a German aviator who spoke in a sort of Ludwig Von Drake accent and piloted his aircraft with raspberries painted on the fuselage. Sir Grapefellow was his British adversary, dressed in purple and sporting grape artwork on his triplane. Like Count Chocula and Franken Berry, these two doughty airmen battled in air, land, and sea over the merits of their respective cereals, with neither coming out a clear victor. It must be said, however, in choosing grape and raspberry for the flavors of its new cereals, General Mills was certainly venturing into territory where no other company had gone—and judging from the two cereals' brief shelf life, there had probably been a reason for that.

In the early 1970s, the Cheerios Kid was still around, but under the pressures of new regulations, he was a mere wisp of his former studly self. One of the NAB rules adopted seemed specifically aimed to take away much of that particular mascot's selling power:

> Any representation of a child's concept of himself/herself or of his/her relationship to others must be constructively handled. When self-concept claims are employed, the role of the product/service in affecting such benefits as strength, growth, physical prowess and growing up must accurately reflect documented evidence.

In plain English, it mattered not that characters from Popeye in the 1930s to Woody Woodpecker in the 1950s to a host of others in the 1960s had been gaining superhuman strength from various cereal brands; under the watchful eyes of the new watchdogs, such exaggeration would not be tolerated. The Cheerios Kid and Sue continued to have rather lame adventures, but now eating Cheerios did not seem to have the same effect on our young friend. He still performed mild stunts, but no longer flexed his Cheerio-decorated bicep to demonstrate the cereal's potency. Sue continued to be a helpless victim, and who could blame her, since eating Cheerios did not really seem to help matters? It was not long before the two youngsters were forced into early retirement, and Cheerios ads were given over to images of live-action families frolicking together, powered by the energy (if not strength) generated by a Cheerios breakfast.

The staple brands Cheerios, Lucky Charms, and Cocoa Puffs continued to sell, even as the world of advertising changed, but during the rest of the 1970s, when it came to introducing new products, General Mills appeared to have lost whatever four-leaf clovers Lucky the Leprechaun had been hiding around the place. First, there was its discovery that cereal pieces could be impregnated with a filling of another flavor entirely. Such was reason enough for the launch of Mr. Wonderfull's Surprize (1972), a cereal that must have made spelling teachers climb right up the chalkboard. Logo character Mr. Wonderfull looked like a slightly less mentally stable Willy Wonka with a periscope hat. Then there was Crazy Cow (1974), with a mascot of the same name, which came in chocolate and strawberry flavors. That by itself would have rendered it pointless from the same company already manufacturing Count Chocula and Franken Berry, but Crazy Cow the character, wearing a crown-like beanie of the type popularized by George "Goober" Lindsey on *The Andy Griffith Show* or Jughead in the *Archie* comics, did not have enough personality to pull the cereal out of its rut. And it did not even begin to answer the question of how a cow could be male.

Magic Puffs (1974) were promoted by the anthropomorphic Magic Hat, a magician's topper with facial features. And then there was the attempted 1977 reintroduction of Frosty O's under the new name Frosted O's. Tying in with the oat theme, Frosted O's used a nameless, bucktoothed horse who was soon put out to pasture. (In the 1990s, Frosty O's were re-reintroduced as Frosted Cheerios, with no logo character, under which name they are still on the market.)

Fortunately, General Mills still had its old dependables, including the Trix Rabbit. Even those veterans were not immune from modern cross-examination; as historian Jim Hall reported, in the late 1970s, the long-running "Trix are for kids" theme came under scrutiny. He wrote, "Perhaps realizing that this parable might teach a damaging lesson—try as you may, you will never achieve your goal—General Mills found a way out of the situation." The quick fix took advantage of 1976 being a presidential election year. Trix boxes were printed with "ballots" that the too-young-to-vote crowd could mail in, choosing whether the rabbit should finally get to eat Trix, or conversely, voting that no, Trix was still for kids only. Hall described the outcome of this particular election:

> Over 99 percent of the ballots were pro. As balloons fell and a brass band played, the Rabbit finally tasted his Holy Grail and went berserk with ecstasy. Then, like Oliver Twist, he held out his empty bowl and asked for more—only to be told to wait until the next election.

The success of this campaign must have warmed General Mills' corporate heart. Newcomers Mr. Wonderfull, Crazy Cow, and Magic Hat might have been total losers, but it was comforting to know the reliable old Trix Rabbit and his cohorts from the late 1950s and early 1960s still had the magic in a vastly different world.

During the 1976 presidential election, youngsters who were too young to vote for Ford or Carter could instead cast a ballot determining whether the Trix Rabbit would finally get a bowl of his favorite cereal, or if it would continue to be reserved for kids only.

We Are the Freakies

The 1970s Get Weird

The other cereal companies that had boomed alongside Kellogg's and General Mills during the advertising wave of the 1960s were merely trying to tread water once the 1970s arrived. Probably the one able to maintain its speed the most consistently was Quaker Oats. With Cap'n Crunch, Quisp, and Quake holding down the fort, Quaker was ready to launch a new product with the help of Jay Ward.

Quaker's new nutrient-enriched cereal was named King Vitaman. (The FDA would not allow the name to be spelled or pronounced "vitamin" since it was a cereal and not a drug.) Naturally, such a name called for a regal figure as its mascot, but in true Ward fashion, the cartoon King Vitaman was not some Prince Valiant–type hero. Ward's king was about three feet tall and spoke with the whining voice of well-known TV and movie wimp Joe Flynn. Other voices in the commercials were supplied by Ward's A team of Paul Frees, Daws Butler, and Bill Scott.

Runty as he might have been, King Vitaman was fearless when it came to protecting his horde of cereal, which was "shaped like little golden crowns." The villains who tried to storm his castle walls included the Blue Baron and the Not-So-Bright-Knight, but the picayune potentate always managed to outwit these halfwits. He had the not-so-stalwart help of

King Vitaman was another Quaker Oats product that had the good fortune to be the subject of a Jay Ward ad campaign. The runty ruler was fearless in protecting his cereal stash from no-goodniks such as the Not-So-Bright Knight.

knights Sir Laugh-It-Up and Sir Craven, but in the end, the meanies would be defeated and everyone would sit down to a balanced breakfast. The standard closing gag was when the ruler would declare, "Now, pour the milk on your King Vitaman," and receive a pitcher of milk dumped over his head. "Not ME, you nincompoops, the CEREAL!" he would scream in a hissy fit.

Ward's King Vitaman adventures were unfortunately short-lived, and by 1973, the fussy figurehead had been replaced by a live-action King Vitaman played by veteran actor George Mann. These spots were played for charm instead of humor, as the kindly king with a crown made of spoons would gently interact with some of the cutest children ever seen in commercials. Even long after his death in 1977, Mann's smiling face graced the King Vitaman boxes until a new cartoon king (looking nothing like Ward's original) was designed in the 1990s.

Meanwhile, back on Planet Q, Quisp and Quake continued to upstage each other at every opportunity. Quake brought in additional ammunition by introducing his ally, Simon the Quangaroo, an orange marsupial who hyperactively bounced around the scenery on behalf of a new product, Quake's Orange Quangaroos. (Simon's voice was Bill Scott doing his best Cary Grant impersonation.) Simon came not from Australia but from Orangeania, which managed to irritate even the usually cheerful Quisp with its monochromatic color scheme.

The Quisp/Quake feud had been going on for so long, with no end in sight, that Quaker hit upon the same solution General Mills would use with the Trix Rabbit four years later. During the 1972 presidential election, Quisp and Quake each solicited votes of support from the kids at home. They solemnly agreed that the winner's cereal would remain on the market, while the loser would disappear forever. Since, to be honest, a blindfolded person in a taste test would probably not have been able to tell the two cereals apart, this was basically a popularity contest for the characters rather than a verdict on quality. The results were revealed in what must have been one of the strangest animated commercials ever scripted, when Quisp won handily and Quake slunk off into nowhere with the agony of defeat weighing heavily on his once-broad shoulders. The aftershocks of this election continue even to the present day. Quisp was eventually discontinued later in the 1970s, but in recent years, Quaker has reintroduced it for limited periods of time. Sticking with their 1972 agreement, however, Quake has made no such reappearances, and one must assume he is living out his old age in his secret lair at the earth's core. Let us hope at least Mother Lode is still there to keep the retired superhero company.

As for Quaker's big gun, Cap'n Crunch, once Crunch Berries proved to be a success, the line continued to grow and mutate into every conceivable flavor. The first spin-off of the 1970s was Peanut Butter Crunch, in honor of which the Cap'n adopted a huge tan elephant named Smedley. This was one instance of Jay Ward actually recycling one of his earlier creations. Ward's final cartoon series for Saturday mornings, *George of the Jungle* (1967), had featured the title character's pet elephant Shep, who thought he was a dog and behaved accordingly. Most of Shep's personality, if not his appearance, was carried over to Smedley, for whom Bill Scott provided the same enthusiastic trumpeting noises.

Even the old miscreant Jean LaFoote the Pirate ended up with his own cereal. Cinnamon Crunch was his accidental invention, and its commercials were unique in that LaFoote spent most of his screen time in them trying to destroy the inventory of his own product, ostensibly so no one else could enjoy it. In a typical gag, LaFoote holds a lighted stick of dynamite over the ship's hold full of Cinnamon Crunch, as he tries to remember, "Now, I throw thees in five seconds . . . or ees eet ten seconds? (BOOM) Eet was five seconds . . ." Cap'n Crunch would ask, "How could

The Cap'n Crunch cast of the 1970s supplemented the original crew with such unlikely allies as Smedley the peanut butter–loving elephant and the biologically indeterminate Crunch Berry Beast.

such a bad guy come up with such a good cereal?," to which the battered LaFoote would protest, "I didn't MEAN to!"

New flavors just kept coming over the next ten years. Vanilly Crunch was introduced by one of Cap'n Crunch's friends from the briny deep, Wilma the White Whale, who seemed to have unhealthy interspecies romantic feelings toward the Cap'n. Punch Crunch tasted just as it sounded and was represented by sailor-suited Harry Hippo, a friendly but clumsy clod whom Bill Scott played in a slight variation of his Bullwinkle voice. Finally, there was Choco Crunch, for which Cap'n Crunch and his crew had to outsmart the Chockle Blob, a morphing monster who tried to swipe the chocolate-chip treat while rumbling, "chockle, chockle, chockle." Most of these supplemental flavors came and went within a year or two, but Crunch Berries and Peanut Butter Crunch (along with the original, of course) proved to be the long-distance runners, still available on grocery shelves today.

In contrast to Quaker's ever-expanding line, Post seemed to retreat a bit after spending most of the previous decade with the huge *Linus the Lionhearted* cast. As the 1970s got under way, Post seemed to be actively trying to downplay the few cartoon mascots it had left. For example, Lovable Truly was reduced to a cameo appearance in his own Alpha-Bits commercials, which were increasingly devoted to crowds of live-action children. By 1971, the steadfast postman was part of the dead letter office.

By the time of this early-1970s magazine ad, Lovable Truly had hung up his mail sack, and Rory Raccoon and So-Hi were standing in the unemployment line. Sugar Bear and Linus the Lionhearted still held down their respective positions, even as the ad copy acknowledged the silliness of their ad campaigns and emphasized the new concentration on nutrition.

His place on the box was first occupied by a gang of youngsters in their psychedelic flying boat, an oh-so-mod element introduced into the commercials. Showing how attitudes were changing, the kids in the ads would swoop down upon a clueless adult and ask him inane questions, then laugh when the old fuddy-duddy was unable to give the proper punch line. Their victim would fume as the young rebels sailed away into the sky.

After the flying boat was grounded, the Alpha-Bits commercials became simpler. A boy wearing an aviator's cap and goggles, for no stated reason, would spin yarns based on the words he was able to spell with his cereal. "Once, a big A-P-E tried to grab me," he would solemnly state. "What did you do?" his female companion would gasp breathlessly, prompting the inevitable: "I ate him." The routine was no "Who's on First?" in longevity, but cartoon art of the goggles boy and the girl would appear on Alpha-Bits boxes for another decade.

There was a similar approach taken with Honeycomb, once the Honeycomb Kid had headed for the last roundup. Now the commercials involved live kids in their "Honeycomb Hideout," sampling their cereal to any number of oversized visitors (a wrestler, a roller-derby queen, a circus

Animation historian Jerry Beck discovered this partly animated pilot commercial for Post's new frosted flakes, which were apparently intended to go by the rather unappetizing name Pink Panther Food.

strongman) who dropped by looking for "something with a really big taste." One of the kids, wearing 1970s headgear looking like a reject from the *Gilligan's Island* wardrobe department, remained on the Honeycomb boxes long after such fashions had gone out of fashion.

Sugar Sparkled Flakes had been taken off the market for a few years by the time Post introduced their replacement in 1972: Pink Panther Flakes, represented by the strawberry-colored feline created by animators David DePatie and Friz Freleng. (Recently animation historian Jerry Beck discovered a 16-millimeter print of an experimental commercial referring to Post's new flakes as "Pink Panther Food." Presumably the name was changed so it would not sound like something one would look for in the pet aisle.)

As might be expected, the commercials for Pink Panther Flakes used a jingle set to the immortal Henry Mancini theme song and depicted the ever-silent Panther trying to sneak a snack from under the huge nose of the nameless little man who was his victim in numerous cartoon shorts. Pink Panther Flakes lasted only a couple of years, and eventually Sugar Sparkled Flakes made a low-key return in limited markets.

When Pink Panther Flakes made it onto the market in 1972, the amusing commercials depicted the magenta feline obtaining a balanced breakfast by hook or by crook, usually from the nameless little man who so often appeared in his cartoons.

One of Post's only original animated characters left in the stable was Sugar Bear, and even he had been affected by the same regulations that had neutered the Cheerios Kid. Giving up all pretense of super strength, the cool crooner partnered in 1972 with his girlfriend Honey Bear and formed a rock 'n' roll group (the Sugar Bears, what else?). Their singing voices were provided by professional musicians, and Gerry Matthews did not have to sing in Bing Crosby's voice any more, although he was kept on hand for Sugar Bear's speaking roles.

By the mid-1970s, the band was disbanded and a new series of commercials began depicting Sugar Bear trying to keep his Super Sugar Crisp out of the hands of various would-be thieves. Adhering to the regulations, he did so not by physical force but by his wits. The villain would invariably overhear Sugar Bear bragging about Super Sugar Crisp's "golden coat" and assume what was in the box had real financial value. This served to set up Sugar Bear's new catchphrase, "It's not a treasure, it's a treat!" The most determined of the bad guys was a Mr. Fox, whose Dixie accent made him a hybrid of Brer Fox from Disney's *Song of the South* and the characters in Walt Kelly's comic strip *Pogo*.

In the Fruity and Cocoa Pebbles commercials, Barney was still tricking Fred out of his breakfast in every contest. One temporary setback was the June 1977 death of Alan Reed, the veteran voice of Fred Flintstone. For approximately a year, Fred's voice was supplied by Reed's son, Alan Jr., and while the imitation was close, it still sounded like Fred was a trifle "off-key." Late in 1978, Hanna-Barbera introduced the new official Fred, Henry Corden, whose interpretation was much closer to the original. Mel Blanc continued as Barney, and the commercials proceeded as if nothing had happened.

In an attempt to change things up from the usual routine, sometimes new themes were introduced into the Pebbles commercials. One of the worst ideas was a short-lived period in the late 1970s when the sight of Fruity Pebbles would transform friendly old Fred into a fanged, hairy "Fruity Monster," while a glimpse of Cocoa Pebbles had the same effect on buddy Barney. Only a spoonful of each cereal, quickly supplied by some live-action kids, would avert the horror that seemed sure to ensue. Fortunately, this ill-conceived series did not last long, and soon the characters were back in their habitual Stone Age setting with Barney's good-natured trickery intact.

Did you notice we have not heard from Nabisco since the late 1950s, with its spoon-sitting Spoonmen promoting Shredded Wheat Juniors? The company just could not seem to get out of a rut when it came to the presweetened cereal market, and that rut was filled with Wheat and Rice Honeys. Whenever Nabisco got away from those, the rutted road led only to oblivion. In 1966, Nabisco had tried giving Wheat Honeys a caramel coat and marketed them under the brand name Puppets. Such a moniker would not have made much sense, except Puppets cereal was packaged inside plastic replicas of Mickey Mouse, Donald Duck, Winnie-the-Pooh, and Kanga and Roo. These figures, which bore a strong family resemblance to Colgate-Palmolive's Soaky bubble-bath toys (both originated with the same premium design firm), were most likely supposed to be used as puppets when empty—but that was not enough to give Puppets a helping hand.

In the early 1970s, Nabisco test-marketed OOOBopperoos, a blueberry cereal promoted by the sunglasses-wearing Blue Kangaroo, and Norman's, another also-ran product that at least had the distinction of Arnold Stang as the voice of its titular mascot character. Those test markets indicated no huge demand for either product, so the Blue Kangaroo was sent back to the Australian outback and Stang went on to become Snappy the Turtle for Kellogg's. Nabisco found itself back at the old drawing board, with Rice and Wheat Honeys as its only friends.

Deciding that if the dual Honeys were all they could market successfully, they would stick with it, Nabisco reinvented the flagship product in 1973. For all of its unsuccessful launches, Nabisco did have one thing going for it: a cozy relationship with the Walt Disney Studios. Throughout the 1960s, Rice and Wheat Honeys had been the outlets for any number of now highly prized Disney premiums, including a set of plastic Winnie-the-Pooh character spoon-sitters of the same general fashion as the Spoonmen. The Pooh toys emerged from the Hundred Acre Wood whenever Disney released a new Pooh cartoon (first in 1966 and again in 1968), so someone with more than a very little brain came up with the idea of making Pooh a true product mascot.

Thus, in 1973, Wheat and Rice Honeys were reintroduced for the eleventeenth time as Winnie-the-Pooh Great Honey Crunchers. Although Pooh had proven appeal, and a new animated film was scheduled for release in 1974, this time the Disney magic did not materialize in the form of sales. Pooh went back to work with Christopher Robin, and Nabisco was stuck with trying to sell Honeys yet another way. This time, they became Klondike Pete's Crunchy Nuggets, represented by an old prospector (Klondike Pete, natch) and his donkey, who was more interested in reading books than helping his owner find gold or presweetened wheat puffs. For the one year Crunchy Nuggets tried to strike it rich, this pair appeared in commercials and numerous premiums, but Nabisco decided it knew when it was licked. It gave up the presweetened cereal business to concentrate on Shredded Wheat and the multitude of other Nabisco products and was glad to be rid of it.

If any company could be said to have emerged in the 1970s to give serious competition to Kellogg's, Post, and General Mills in sheer number of new cereals and their matching mascots, it would have to be the dark horse candidate of Ralston. There was something of a false start in 1971, when Ralston decided to supplement Rice, Corn, and Wheat Chex with a new children's-oriented Sugar Chex. For no particular reason, Ralston signed up Casper the Friendly Ghost as the mascot for Sugar Chex—and with a dead character on the box, it was not long before Sugar Chex followed him to the cereal graveyard.

For the rest of the decade, Ralston fell into a pattern of introducing a new product, with new characters, each year. Much advertising would be thrown behind each one, but the inescapable fact was they all tasted alike, and even that taste was not original: it was Cap'n Crunch's, which must have been the easiest to duplicate in the whole cereal industry. As happens so often, the first cereal in this new Ralston parade turned out to be the most memorable.

Only a desperate company would have thought of calling its new cereal Freakies, and only a despondent company would have agreed to mascots Scott Bruce described as looking "like Disney's Seven Dwarfs after an encounter with a toxic waste dump." That might be taking things a bit to the extreme, but there were indeed seven Freakies in the cast, perhaps deliberately inviting such comparison. According to Bruce, the characters were the creation of ad executive Jackie End, who secretly modeled their personalities off those of her coworkers at the Wells, Rich, Greene Agency (obviously the public was unaware of the satire, and with each Freakie having so little screen time in the commercials, it is difficult to say just how far End's parodies actually went).

Probably the main reason Freakies is so well remembered by late-era baby boomers is its two commercials that ran incessantly on Saturday-morning television during 1974–75. One described how the Freakies were wandering through the world, lost and alone, with no place to call home. The narrator explains they were fortunate to have a leader like Boss Moss to run to for help. "We're scared!" they wail while clinging to Boss Moss's warty hide. "I understand exactly how you f-f-feel," he stammers (yes, Jackie End named Boss Moss after her own boss, Charlie Moss). Finally, parting some foliage, the Freakies discover a magical tree dripping with

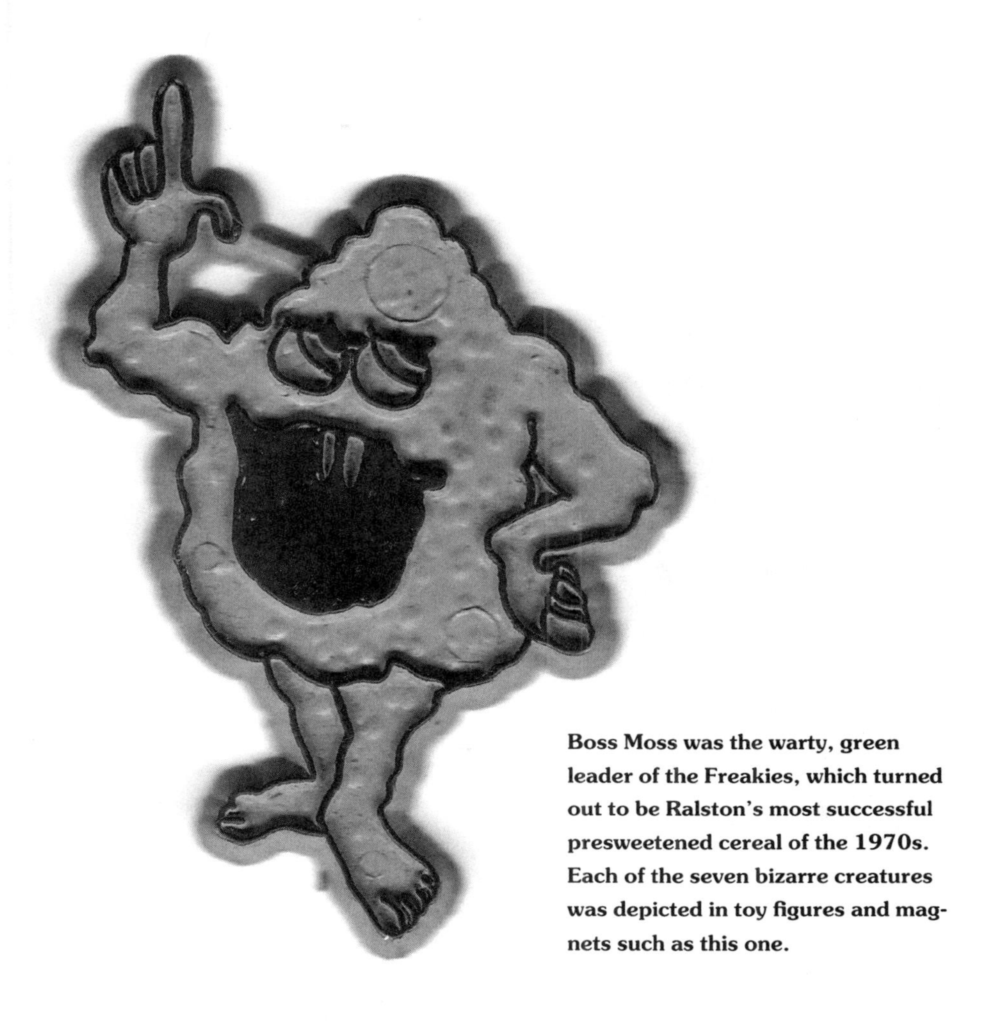

Boss Moss was the warty, green leader of the Freakies, which turned out to be Ralston's most successful presweetened cereal of the 1970s. Each of the seven bizarre creatures was depicted in toy figures and magnets such as this one.

"their favorite cereal, Freakies," according to the narrator, leaving us all to wonder where the cereal got its name if it existed before the seven Freakies found the tree.

The other commercial was a musical one, involving the whole cast in a song-and-dance number. In case you want to know, the six other Freakies besides Boss Moss were Gargle, Grumble, Goody-Goody, Snorkeldorf, Hamhose, and Cowmumble. Each introduced himself/herself/itself during the song, which was set to the tune of the old 1899 chestnut "Hello Ma Baby," a.k.a. "Telephone Rag." To this melody, the gang of plug-uglies sang, "We are the Freakies, we are the Freakies, and this is our Freakies Tree; we never miss a meal, 'cause we love our cer-e-eel. . . ."

True to Ralston's modus operandi for the decade, the Freakies were cast back out into the wilderness in 1975, and another similar-tasting cereal was brought in to fill the void. In fact, according to its backstory, the

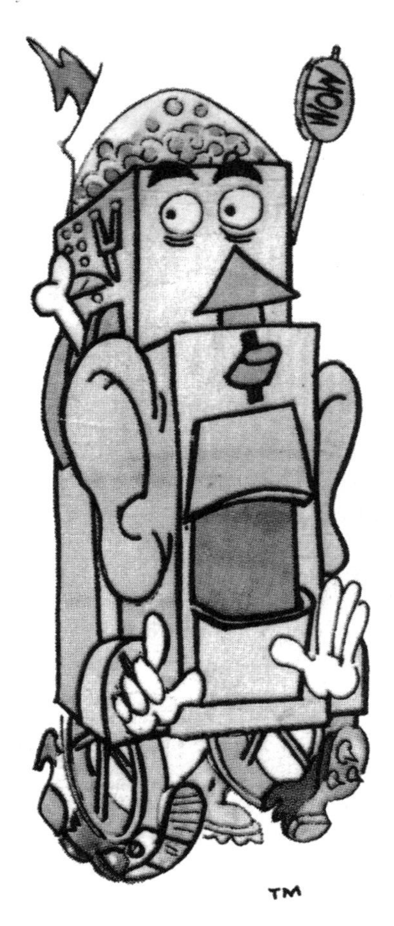

Cecil the Computer had a voice and personality modeled after golden age of radio comedian Fred Allen. Cecil grudgingly produced mass quantities of Ralston's new cereal Grins & Smiles & Giggles & Laughs in 1976.

cereal *originated* in a void. Scott Bruce gave the most concise synopsis: "Moonstones were manufactured under the lunar surface by Moonbeams, a noble race led by Major Moon . . . Out to steal their secret formula were villains from the dark side, called Moonbums. Their leader was Bigbum, 'the biggest bum of all.'" In truth, this sounds like the type of plot that might have been cooked up by a group of seventh-graders, but it did not really matter since after its customary one-year shelf life, Moonstones became just another bunch of space junk.

The 1976 cereal had an even stranger name, with a heavy overload of ampersands: Grins & Smiles & Giggles & Laughs. Those were also the four human characters in the animated commercials: three males and one female (the lady, of course, was Giggles). The true star of their adventures was a sentient computer named Cecil, who had at least one characteristic setting him apart from all other mascots. His voice was an imitation of dour 1930s and 1940s radio comedian Fred Allen, whom few people could have been expected to remember in the days before recordings of radio's golden age shows became readily available.

The general idea in the G&S&G&L commercials was that if Cecil were entertained with sufficient comedy material, he would produce a heaping helping of the cereal. If the quartet's jokes fell flat, he would give them only a pie in the face. In true comedy fashion, usually the staff's intended boffo gags would fail to amuse Cecil, but an accidental pratfall on their part would bring forth a torrent of the cereal—not loose pieces, but already-boxed packages of the stuff. Since it was shaped like miniature smiley faces, G&S&G&L was touted as "the cereal that smiles back at you."

Ralston may or may not have had any greater hopes for its 1977 product, but for one reason or another it went on to become to most successful of all—as proven by the fact it is the only one of these cereals to still be on the market. Cookie-Crisp came in two distinct flavors, vanilla wafer and chocolate chip, which at least distinguished it from all the other Cap'n Crunch taste-alikes from Ralston. The original mascot for Cookie-Crisp was Cookie Jarvis, a wizard with a beard that looked as if someone had glued a box of cotton balls onto his chin. His name was derived from his habit of fluttering around breakfast tables, touching cereal bowls with his magic wand and changing them into cookie jars (yes, it sounds like those seventh-graders were at work again). On one of the first boxes, the miniature magician explained the concept of this new experience in breakfast:

> Kanoodle, Kazaam and how do you do? I'm Cookie Jarvis to tell you about a delicious cereal we've made for you! Cookies for breakfast, you say? Well heavens no, unless they're made of cereal and part of a complete breakfast to help you grow. All around the country kids are saying "cookies for breakfast," that's something new, but kids in America aren't scared of things new. They know that if Christopher Columbus had not tried something new, people would still think the earth is square!

With Cookie-Crisp raking in the chips—the chocolate kind, that is—Ralston ended the 1970s by dragging out an idea that had not worked before, so there was no reason to think it would not work again . . . right? You will recall back in 1966 the company had tried Mr. Waffles as its first presweetened treat. In 1979, what amounted to the same product

Cookie-Crisp, with its magician mascot Cookie Jarvis, was introduced in 1977. It is the only one of Ralston's many pre-sweetened experiments to still be on the market, although it is now manufactured by General Mills.

appeared again, this time named Waffelos and sporting a different character, the Yosemite Sam–like Waffelo Bill. Waffelos came in both regular and blueberry flavor, but even though their popularity waffled back and forth for a couple of years, they were eventually mired in their own sticky, syrupy sweetness.

It may be that one reason new cereals from any company had trouble catching on during the 1970s was that the decade just felt different. The last of the baby boom children were entering high school, and there were not as many children coming behind them to sustain the momentum of the 1950s and 1960s. Much has been written about how the entire country seemed disillusioned during the 1970s, what with a disgraced president, skyrocketing fuel prices, and an unwinnable war in Southeast Asia. Questioning authority was the new craze, and one of the many ways it manifested itself was in a legendary series of bubble-gum stickers known as Wacky Packages. In case you were not a kid yourself during that time, Wacky Packages looked much like something out of *Mad* magazine, with popular household products mercilessly parodied in full-color artwork duplicating their actual logos so closely that many adults who were not in on the gag did not even notice the spoofery. The companies whose trademarks were being mocked certainly noticed, and their lawyers let the Topps bubble-gum company know about it. However, since parody is protected as part of free speech, they were unable to do much except fume impotently. Topps, in any case, was more interested in making peace than war, so for the most part chose to honor any cease and desist letters it received from Kellogg's, General Mills, et al.

The cereals we have seen so far certainly came in for their fair share of ribbing during the years the Wacky Packages art was being produced. Kellogg's single-serving Snack-Pack became, in Topps's version, "Kooloff's Snatch-a-Pack," picturing a sweating Tony the Tiger on "Frosty Snow Flakes," a moronic Ogg the Caveman on "Cocoa Krusties," and a demented fruit man on "Apple Jerks." General Mills was caught in the line of satire when Trix became "Triks, the Practical Joker Cereal," with an illustration of the rabbit pouring hot sauce over a bowl of the stuff instead of milk. Post had to grit its teeth as Super Sugar Crisp became "Super Cigar Crisp," with a stogey-chewing "Stupid Bear" advising kids to "puff your way through breakfast." And then Alpha-Bits turned up as "Awful Bits," with a scowling version of Lovable Truly delivering oat-shaped letters spelling out E-C-H. Quaker was surely quaking over "Cap'n Crud," which "tastes cruddy, even in milk."

Wacky Packages became one of the hottest 1970s fads and inspired a host of loyal fans that still document their history through books and websites today. This can probably be chalked up to the generally pessimistic outlook of their period. Fear not, though, because the 1980s were coming, and things were about to change. Was that for better or for worse? It depends on your point of view, but in our final chapter we shall try to make some sense of it all.

How About a Vitamin-Packed Punch?

The Baby Boomers Grow Up

On our last visit with Kellogg's, Sugar Corn Pops' mascot Big Yella had just started pushing up daisies. His replacement in 1980 was the first Sugar Pops character in the brand's history to not be connected with a western theme: Poppy the Porcupine. This was perhaps the most androgynous character the Leo Burnett agency ever created: to this day, cereal historians seem to be unsure whether Poppy was male or female (if she were the latter, it would make her one of only a small number of female mascots). Poppy's voice could have been mistaken for either, and Poppy's bushy mane of quills certainly reminded one of a typical early 1980s "big hair" look.

Like the other companies we have seen, Kellogg's found sheer market saturation made it difficult to launch any new cereals with true sticking power. In 1985, the company tried putting out OJ's, which coated its puffs with genuine orange juice. Since the Whippersnapper and Big Yella had already gone on to be ghost riders in the sky, Kellogg's picked another cowboy to represent the new product: OJ Joe. In the well-animated commercials, Joe herded live-action rolling oranges as if they were dogies, using his ink-stamper branding iron to emblazon each one with his OJ

Kellogg's newest character, Poppy the Porcupine, arrived just in time to join the rest of the gang in this 1980 coloring book celebrating the company's seventy-fifth anniversary.

monogram. As with most of the other experimental cereals of the era, OJ's hung around for approximately one year before running out of juice.

We shall see this trend again, but during the 1980s companies began renaming their cereals to get rid of that dreaded word "sugar," which had come to represent all the criticism directed toward them by the consumer groups. In short order, Sugar Corn Pops became simply Corn Pops, and next to fall was Sugar Smacks. In 1986, the sweetened wheat puffs became Honey Smacks, and there was a feeling that the new name called for a more appropriate logo than a hyperactive frog leaping about and shouting, "Dig'em, dig'em." Instead, the new Honey Smacks boxes and commercials introduced the public to Wally the Bear, whom Kellogg's felt would emphasize the cereal's honey content rather than its sugar rush. Dig'em had left some awfully big tennis shoes to fill, though, and Wally's flat feet just were not the right size. After a year, Wally was out and Dig'em triumphantly hopped back into place as if he had never left. The active amphibian is still featured on the Smacks boxes today, although in recent years his original early 1970s "cool" wardrobe has been updated to make him better fit in with modern teen culture.

What do you get when you take at least six different cereals and combine them (plus raisins) into a single product? If it happened in your own kitchen, you might call it a mess, but in 1990, Kellogg's did the same thing and called it Bigg Mixx. The mascot character shared the same name—the first time Kellogg's had followed the precedent set by Cap'n Crunch—but a more unlikely figure could hardly be found in the cereal aisle. Bigg Mixx the character was a beast combining the physical features of a moose (antlers), a chicken (comb), a pig (snout), and wolf (body), plus maybe a few others yet unknown to biologists. Neither the character nor the cereal was a hit, so the mixed-up mongrel crept back into the forest.

Before leaving Kellogg's in the early 1990s, it should be mentioned that around that time, Cocoa Krispies brought back at least some version of the monkey who had monkeyed around with it back in the late 1950s. You may remember that we said there was a dispute over whether that simian's name was Jose or Coco. Well, his 1990 reincarnation was definitely named Coco but existed in two separate forms: as a typical cartoon monkey and also for a while as a live chimp. Neither one lasted long, and like Frosted Rice, Cocoa Krispies was eventually folded into the whole Rice Krispies family, where Snap!, Crackle!, and Pop! are still standing guard as they have for some eighty years.

Kellogg's had never been greatly affected by the early 1970s regulations about cereal commercials making exaggerated claims of strength and so forth. As we have seen, though, Post and General Mills had found it necessary to alter or completely eliminate some of their most recognized characters' routines. Well, in the early to mid-1980s, the pendulum began to swing back in the other direction, and it all had to do with the changing political climate in the United States.

After Ronald Reagan became president in January 1981, suddenly one of the hot issues was "deregulation." Shackles that had been placed on large corporations for years began falling away as if by Lucky the Leprechaun's magic, and how one felt about such developments depended largely on one's point of view. A federal court judge ruled in March 1982 that the National Association of Broadcasters code was "an illegal restraint of trade." The advertisers, of course, reveled in their newfound freedom while the reform groups such as Action for Children's Television could only stand by, weeping, wailing, and gnashing their teeth. For good or bad, the 1980s became the decade when toy companies turned their product lines into half-hour commercials; it is somewhat surprising that no cereal company attempted to do what *Linus the Lionhearted* had accomplished, but as it was, they were kept pretty busy just figuring out how to make the new deregulations work best for them.

Post was one of the first to get things cranked up again. The kids who were watching cartoons in the early 1980s would not even have remembered when Sugar Bear used to eat a bowl of Super Sugar Crisp to gain instant strength for rescuing Granny Goodwitch. A new campaign begun around 1981 took the core idea but expanded it to what some might consider an extreme. Now Sugar Bear was redesigned to look like he spent most of his non-commercial time working out at the gym, and Gerry Matthews's Bing Crosby voice was silenced. The usually mild-mannered bear was pictured enjoying breakfast with various groups of anonymous kids when some sort of danger would threaten or a villain would try to make off with the cereal. Roaring, "This is a job for SUPER BEAR," the veteran mascot would explode out of his blue sweater and transform into an Incredible Hulk–like superhero. A few swipes with his clawed fist—or a growl capable of shaking the foundations of the earth—would be all it took to dispatch said villain to the great beyond, and Sugar Bear would return to his normally buff appearance.

The "Super Bear" campaign did not last very long, and soon Gerry Matthews was back in the recording booth. One change turned out to be permanent: with so many other products eliminating the word "sugar" from their names, Super Sugar Crisp became Super Golden Crisp and, later, simply Golden Crisp. It says something about its trademark's beloved reputation, though, that Sugar Bear retained his name on the front of his shirt well into the twenty-first century, even though modern-day youngsters would make no connection between it and the product he represents. ("Daddy, why isn't he named Golden Bear?") In recent years, even that monogram has been reduced to the simple initials "S.B.," heading off any further interrogation.

The new commercials were even more reminiscent of the original pre-regulation versions from the 1960s. Once again, an imperturbable Sugar Bear would be faced with all manner of menaces, from raging bulls to hungry alligators, that would try to strong-arm him out of his cereal. Gulping down a few wheat puffs, Sugar Bear would reply, "How about a vitamin-packed punch?" and deck the villain. Even the old 1960s jingle was brought back, with updated lyrics to match the new name: "Can't get enough of Super Golden Crisp—it's got the crunch with punch!"

In 1987, one of the Super Golden Crisp premiums was a small stuffed figure of Sugar Bear (which, frankly, did not resemble the character nearly as closely as some other items offered over the years). Kids who sent in the requisite box tops for it also received a small postcard that, somewhat unusually, promoted an upcoming commercial as if it were a new movie:

> Hello, this is your pal Sugar Bear! Thanks for writing for your very own Sugar Bear stuffed toy. Now we can go on exciting adventures together! Look for me on TV in my safari adventure, and watch me spin the stripes off three cereal-craving tigers! Well, time to go! Can't get enough of Super Golden Crisp—it's got the crunch with punch!

Above Sugar Bear's printed name at the bottom of the letter was the image of a paw print. Do you suppose this was supposed to indicate that longtime advertising icon Sugar Bear was, in fact, illiterate?

Also in 1987, Post made a huge push toward reviving one of its long-dead products. Crispy Critters was brought back to life, but for some reason, Linus was no longer offered the job of mascot. (Post had continued to sponsor the Linus balloon in the Macy's Thanksgiving Day Parade up through 1984, long after few of the viewers would have even remembered who he was.) Instead, the new representative for Crispy Critters was a Muppet-like creature named Crispy; in the long-established tradition of basing characters on celebrity voices, Crispy was a takeoff on Jimmy Durante, complete with oversized schnozzola and a voice to match. In his commercials, Crispy's favorite term to describe Crispy Critters was

"indubitably," which gave him ample opportunity to add extra syllables in true Durante style (as the real Durante was once quoted, "I love ta mangle da big woids just ta hear 'em scream").

The new version of Crispy Critters failed to catch on, perhaps because the original cereal animals had been sugar frosted, but the 1980s recipe cut back on that ingredient. There were a number of Crispy premiums, including a pair of Little Golden Books featuring the big-nosed buffoon and his animal-cracker-looking friends, but the new Critters could not survive in the changed world of the advertising jungle. (However, one

This premium set of slides was a frame-by-frame re-creation of one of the late-1980s Super Golden Crisp commercials, in which Sugar Bear used his hidden reserves of strength to all but skin a gang of tough-talking tigers.

of the pals who used to hang out with Linus was still present in image, if nothing else. Rory Raccoon continued to appear on the Sparkled Flakes packaging as late as 1989, although the product was available in very few stores.)

Along with Super Sugar/Golden Crisp, Post's other mainstay was still Fruity and Cocoa Pebbles. The commercials with gullible Fred and crafty Barney plowed full steam ahead throughout the 1980s, with a new twist: they became topical, with veiled references to hit movies or TV shows of the era. Barney would disguise himself as a prehistoric version of Batman

Post made an unsuccessful attempt to revive Crispy Critters in 1987. Instead of Linus the Lionhearted, the new mascot was a Muppet-like creature known as Crispy. His voice and personality were based on the late, great Jimmy Durante.

or Indiana Jones, for example, and dazzle Fred with his star power until something caused his ruse to unravel.

The normal pattern was turned on its head for a special, and memorable, Christmas commercial. Dressed as Santa Claus, Barney manages to con Fred once more, but just as they are getting into a fight over it, the real St. Nick comes down the chimney and reminds them both Christmas is the season for giving. A repentant Fred responds by actually giving

In 1989, former *Linus the Lionhearted* cast member Rory Raccoon was still appearing on boxes of Post's Sparkled Flakes, although the product was no longer easy to find on store shelves.

Barney a bowl of Pebbles, wishing him a heartfelt "Merry Christmas, Barn." (Cartoon historians have for years pondered how the Flintstones could possibly celebrate Christmas when they live thousands of years before the birth of Christ.)

Over at General Mills, the 1980s were starting off with a considerably different flavor than the latter half of the previous decade. Speaking of different flavors, the company's big hit of the era was Honey Nut Cheerios. Such a simple, down-to-earth product called for a mascot diametrically opposed to such 1970s losers as Crazy Cow, Mr. Wonderfull, and Klondike Pete. Honey Nut Cheerios was the domain of a friendly little yellow bee that initially had no name but did speak with Arnold Stang's voice. (Stang, as you may recall, was fresh off his forgettable period as Kellogg's Snappy the Turtle.)

The nameless Honey Nut Bee proved to be one of the company's most enduring and endearing mascots. In 1999, General Mills sponsored a balloon of the little insect in the Macy's parade, where he carried on the tradition begun by Post's long-departed Linus balloon. Shortly thereafter, a contest resulted in the bee finally getting a long-overdue cognomen, and Buzbee (as he is now known) continues to grace the Honey Nut Cheerios boxes well past his thirtieth anniversary. Along the way, his voice characterization was assumed by Billy West, replacing Arnold Stang, but the change has had no appreciable effect on the little stinger's popularity.

General Mills could still have an occasional misstep. In 1981, they attempted to bring back Frosty O's under the name Powdered Donutz. The mascots were simply pieces of the cereal with faces, but their sales pitch fell into a hole. Coincidentally or not, the same year, Ralston introduced virtually the same product under the name Dinky Donuts, but mascot Mr. Dinky (who wore a space suit with a propeller atop his cap) was even less inspiring than the talking Donutz.

There was an interesting, if brief, experiment in the General Mills commercials of the mid-1980s. Someone decided to try teaming some of the famous mascot characters with even better-known cartoon stars of the past. In one commercial, for example, Sonny the Cuckoo decides he can be safe from the kids who tempt him to go cuckoo by taking a sea voyage. No sooner is he sailing the bounding main than he discovers the ship's captain is none other than Popeye the Sailor. Unfortunately for both of them, the kids have stowed away aboard the boat and shove a bowl of Cocoa Puffs under Sonny's beak, causing him to lose control and shipwreck them all on a tropical island.

In an even more unusual concept, the Trix Rabbit is seen on a television screen, ending a typical commercial by once again failing to secure any Trix through one of his disguises. A gloved hand reaches into the scene and turns off the TV, as the camera pulls back to reveal Bugs Bunny as the viewer. Deciding, "Dis rabbit needs help," Bugs calls up his fellow rodent on the phone and states he can help him via his vast knowledge of trickery. Both the Cocoa Puffs and Trix commercials were done as two-part stories, but for some reason, the first half of each is much easier to find among video collectors' archives today. The second halves, showing how Popeye rescues the gang from their island imprisonment and just how Bugs proposes to help the Trix Rabbit, appear to be lost.

In 1986, General Mills teamed up with the Crime Prevention Council for a joint promotion with a "Safety Is No Accident" theme. Besides the expected premiums (a coloring/activity book, decals, and the like), General Mills also sponsored a traveling marionette show that made the rounds of malls and shopping centers across the country. The "Big G Cereal Players," as the cast was called, were large puppet versions of the Trix Rabbit, Count Chocula, Franken Berry, the yet-unnamed Honey Nut Bee, Sonny the Cuckoo, and Lucky the Leprechaun, with the Crime Prevention Council's McGruff the Crime Dog thrown in for good measure (even though Boo Berry appeared in the promotional materials, for some reason or other, he was not part of the marionette group).

The puppet show's story concerned a young boy left at home alone by his parents and the safety hazards he would have surely faced if not for the help of his General Mills friends, who came to life out of the various stuffed toys and other premiums scattered about the living room. The sound track for the show was prerecorded by the same voice actors as the commercials, but pains were taken to not overtly promote any of the specific cereal brands in the script. Except for all the characters being immediately identifiable with their respective products, practically the only reference to their television spots was when Lucky briefly commented that he was "always bein' chased by children."

Immediately after this traveling show's tour, a fourth "Monster Cereal" was added to the General Mills roster. Since Fruit Brute had received a silver bullet several years before, the slot for a fruit-flavored cereal to complement the others' chocolate, strawberry, and blueberry flavors was wide open. So the tomb was opened for the debut of Fruity Yummy Mummy, an Egyptian cadaver wrapped in pink, yellow, and purple bandages. Like his wolfish forerunner, Yummy Mummy did not have a

This activity book was offered in conjunction with General Mills' joint "Safety Is No Accident" campaign with the National Crime Prevention Council. The famed mascot characters temporarily gave up selling cereal to impart worthwhile lessons to kids about living safely in a dangerous world.

celebrity impersonation for a voice, and his reign was just about as brief. At least as of this writing, Yummy Mummy was the company's last attempt at combining thrills with breakfast, although the three original creepy mascots continue to lurk in grocery aisles nationwide.

You may recall back in the early 1970s, one of Quaker's Cap'n Crunch spinoffs had been Jean LaFoote's Cinnamon Crunch. With that recipe long gone and forgotten, General Mills brought back the same basic concept in

Fruity Yummy Mummy, introduced in 1987, was the newest General Mills "Monster Cereal," but the Egyptian carcass in multicolored bandages had wrapped up his short-lived ad campaign by the early 1990s.

1988 as Cinnamon Toast Crunch. Initially the mascots were three bumbling bakers who stumbled over each other like the Three Stooges as they attempted to produce their cereal; the short, fat baker was known as Wendell, with the other two bearing the unmemorable names of Bob and Quello (Quello?). Seeing the three of them together definitely conjured up visions of Snap!, Crackle!, and Pop!, so whether or not Kellogg's had anything to do with it, after a few years Bob and Quello were retired, and Wendell was left with the whole Cinnamon Toast Crunch bakery operation to himself.

The new deregulation that brought back Post's ad campaign with Sugar Bear had an even more dramatic effect on one General Mills product. In the mid- to late 1980s, without any warning, the Cheerios Kid and Sue were back in the commercials, acting as though they had never left. Just as in the old days, dangers would threaten them and a bowl of Cheerios provided the necessary vim, vigor, and vitality to combat the menace. But guess what? Changing times finally caught up with the aggravating double standard that had been in place since the Wheatena radio commercials of the 1930s.

Now, not only did the Cheerios Kid gain super strength from eating his namesake product, but so did Sue. The two youngsters would jointly flex a pair of bulging biceps with the familiar oat circle design, and Sue would jump right into the action to help beat the villain of the day into a helpless pulp. One can almost hear thousands of TV viewers shouting, "It's about time!"

Down at Checkerboard Square, in the early 1980s, Ralston discovered a much easier way to sell cereal than the trial-and-error process of coming up with new mascot characters one after another. The company decided it would be just as profitable to keep packaging basically the same cereals over and over again, while licensing well-known characters from their owners to serve as both the theme and the representative figures of their products. Each of these licenses was set to run for only a short period of time, after which the same product would be assigned a new motif. Using this method, Ralston ate up the '80s with cereals named after the Donkey Kong and Super Mario Brothers video games; movies, including *Gremlins, Ghostbusters,* and *Batman*; beloved toys such as G. I. Joe, the Cabbage Patch Kids, Barbie, and Rainbow Brite; and even other food products, including Cracker Jack, Dunkin' Donuts, and Nerds candy.

Even among all the repetition of these short-lived cereals, Ralston would occasionally get back into the business of its own original creations—but *only* occasionally. Tired of Cookie Jarvis waving his magic

Three bakers named Wendell, Bob, and Quello bungled their way through producing Cinnamon Toast Crunch for General Mills in the late 1980s. Eventually the first two were bought out, and Wendell remained as the sole mascot.

wand about, in 1985, the wacko wizard was replaced in the Cookie-Crisp ads by a true bad guy, the Cookie Crook and his identically masked bulldog, Chip. This unsavory pair was constantly trying to swipe the nearest supply of Cookie-Crisp, only to be arrested by the Keystone Kop–like Cookie Cop.

In 1987, Ralston tried to revive Freakies, but the deformed creatures did not have the same appeal as they had to kids thirteen years earlier. The same year, the company unveiled a new fruity concoction known as Fruit Islands, hosted by a potbellied South Seas chieftain known as King Yummy-Yumma. (Inasmuch as this was the same year General Mills excavated Fruity Yummy Mummy, some perhaps not-so-accidental confusion was bound to occur.) True to Ralston form, Fruit Islands disappeared beneath the waves within a year.

The company's fruit-flavored replacement for Fruit Islands was a little ahead of its time as far as theming was concerned. The big interest in roadside tourism history was still a few years away when Ralston introduced Dinersaurs, complete with the imagery of a roadside diner shaped like a prehistoric beast (as would be so well documented by future researchers, such architecture had been extremely popular in Southern California during the golden age of automobile tourism). This dino diner boasted a comical saurian staff, but like their Ralston predecessors, the Dinersaurs were extinct by 1989.

Ralston's licensing of existing characters to serve as mascots came to its ultimate head in 1989 with the introduction of Morning Funnies, yet another cereal shaped like grinning faces (it seems Cecil the Computer had managed to find another job in the Ralston production line after the demise of Grins & Smiles & Giggles & Laughs). The difference was instead of having one logo character, or even two or three, the Morning Funnies boxes were splashed with actual reprinted newspaper comic strips featuring the stars of *The Family Circus*, *Marvin*, *Tiger*, *Dennis the Menace*, *Hi and Lois*, *Beetle Bailey*, *What a Guy*, and *Luann*. As if the packaging were not already crowded enough with such funny folk, Morning Funnies reached all the way back to the days of General Mills' Twinkles and had a perforated back panel that could open to reveal even more comic strips. This dependence on reading material is probably what made Morning Funnies stop its presses soon afterward; comic strips are fine and funny if read once or even twice, but before the box could be emptied, they had become old and tired and consumers were in no big hurry to buy another box to read their next installments. It must be said, though, while it lasted, Morning Funnies had the largest single cast of mascot characters in cereal history.

After making it through most of the 1990s with still more licensed, temporary characters and cereals, the Ralston pedigree met an unusual fate. In 1997, Ralston's cereal division was absorbed by General Mills, which decided to continue manufacturing the three flavors of Chex (which by now no one associated with the old Checkerboard Square address) and Cookie Crisp (now without a hyphen in the name). For the latter, General Mills jailed the Cookie Crook and retired the Cookie Cop with a well-deserved pension but kept Chip, the masked burglar dog, as the only mascot.

Besides changing market trends, another thing the cereal companies and their respective ad agencies had to deal with during the 1980s was the increasingly frequent demise of many of the voice actors who had been so dependable in their commercials since the 1950s. One of the first to go was Paul Frees, who died on November 2, 1986. He was still recording

Ralston's Morning Funnies, marketed briefly in 1989, outdid all other cereals by having as its mascots the casts of at least eight different newspaper comic strips. As with General Mills' Twinkles in the early 1960s, the back of the box opened up to form a comic book.

new material as Toucan Sam right up to that time, so Kellogg's had to scramble to find a hasty replacement. The role finally went to Maurice LaMarche, who has maintained Frees's Ronald Colman impersonation even long after most people have forgotten the voice was based on a real personality. LaMarche told historian Tim Lawson, "I worked on [Frees's] second-to-last Froot Loops commercial and he was sick at the time. Just for timing, I read Paul's lines for Toucan Sam a few times. They remembered that, so when Paul passed away, I got a good shot at it."

Kellogg's was more fortunate when it came to its other biggest star, Tony the Tiger. Even after a long battle with throat cancer, Thurl Ravenscroft remained the iconic character's voice until his death at age ninety-one on May 22, 2005. His fifty-three continuous years of playing the same mascot set a record in the advertising business that will probably never be equaled, much less surpassed. Robert McFadden, a mainstay of the General Mills commercials (and, earlier, the Post campaigns), had been retired for at least a couple of years at the time of his death on January 7, 2000. The biggest name of all in voices, Mel Blanc, had died in 1989, still playing Barney in the Fruity and Cocoa Pebbles commercials to the end. The role, in all of Barney's animated appearances, was soon assumed by another longtime Hanna-Barbera actor, Frank Welker.

Daws Butler died on May 19, 1988. Cap'n Crunch was the only cereal mascot he was still playing with any regularity, and even his long-running role had presented some difficulties for him. In the early 1980s, Butler suffered a stroke that miraculously did not impair his speech, but affected his brain's ability to process information while reading a script. Frustratingly, Butler found himself skipping words and lines, so his last years of Cap'n Crunch spots were done via a director reading his lines to him and Butler repeating them back in the proper voice. Shortly after Butler's stroke, Jay Ward had given up production of the commercials because the new powers-that-were at Quaker decided they knew more about the business than he did and began dictating what would and would not be included in the scripts.

Ward and Bill Scott had always prided themselves on being the iconoclasts of the animation industry, and when the new social mores of Saturday mornings came along, the delirious duo chafed under the increased regulations. In 1981, Scott vented to TV historian Gary Grossman that they had already had to take away Cap'n Crunch's sword. The plots of the commercials had been sanitized to the point of blandness. "There's no stomping or running," Scott elaborated. "We can't show the flexibility that animated characters are known for. Once we lock ourselves like that,

One of the last ad campaigns Jay Ward's studio launched for Cap'n Crunch was a series involving the ancient mariner's return to his former home, Crunch Island. This comic-strip adaptation of one of the commercials shows that Ward's penchant for visual puns was as sharp as ever.

we might as well be doing live-action programs." It was no longer fun for Ward and his crew, especially after Scott's death in 1985, so Ward reluctantly left the Cap'n in other hands, whether they were capable or not.

It could have been some of the same Quaker ad executives who were responsible for one of the biggest advertising faux pas of the decade, and, coincidentally, it echoed some of the events described in the first chapter of this book. It all began when Quaker Instant Oatmeal needed a new ad campaign, and someone had the brilliant idea of bringing in Popeye to be the new mascot. This was no quick-and-easy, afterthought-type idea; the whole thing was tied together with TV commercials, print ads, comic books to be included in the boxes of instant oatmeal, and an available membership in "Popeye's Quaker Club." It has never been determined whether anyone connected with this new idea had ever been exposed to Wheatena's Popeye radio series of the 1930s, but the similarities were so strong that it seems almost certain some of those old transcriptions were circulating about the agency somewhere.

In his new Quaker adventures, Popeye and his friends would be faced with imminent danger, whether it be from bad guy Bluto, terrors of the jungle, recalcitrant sea creatures, or anything else. Olive Oyl would proffer the traditional spinach can, to which Popeye would yell, "Can the spinach! I wants me Quaker Instant Oatmeal!" Just as Wheatena had worked as an admirable spinach substitute more than fifty years earlier, a quick bowl of instant oatmeal would produce a gargantuan muscle in Popeye's upper arm, glowing with the image of a complete breakfast, and enable him to give the bad guy a whipping in his own traditional cartoon style. Having disposed of his adversary, Popeye would sing new lyrics to his theme song: "I eats me oatmeal and I'm stronger than steel/I'm Popeye the Quaker Man!"

Right there is where Quaker Oats dropped its oatmeal barrel. Apparently no one bothered to remember—or, perhaps were not aware in the first place—that the Quaker Oats Company was named after the virtues of the Quaker church. And, one of the major tenets of that church was its policy of nonviolence. Therefore, having a brawling, muscle-bound mariner proclaim himself as "Popeye the Quaker Man" did not go over too well with the congregation. As we saw way back in chapter 1, the church had not been thrilled when the oatmeal kings appropriated its name in the beginning, and this was just one more blister in a long and contentious relationship.

The Quakers and Quaker Oats sniped at each other in the media for several months, with the latter trying to put the toothpaste back in the

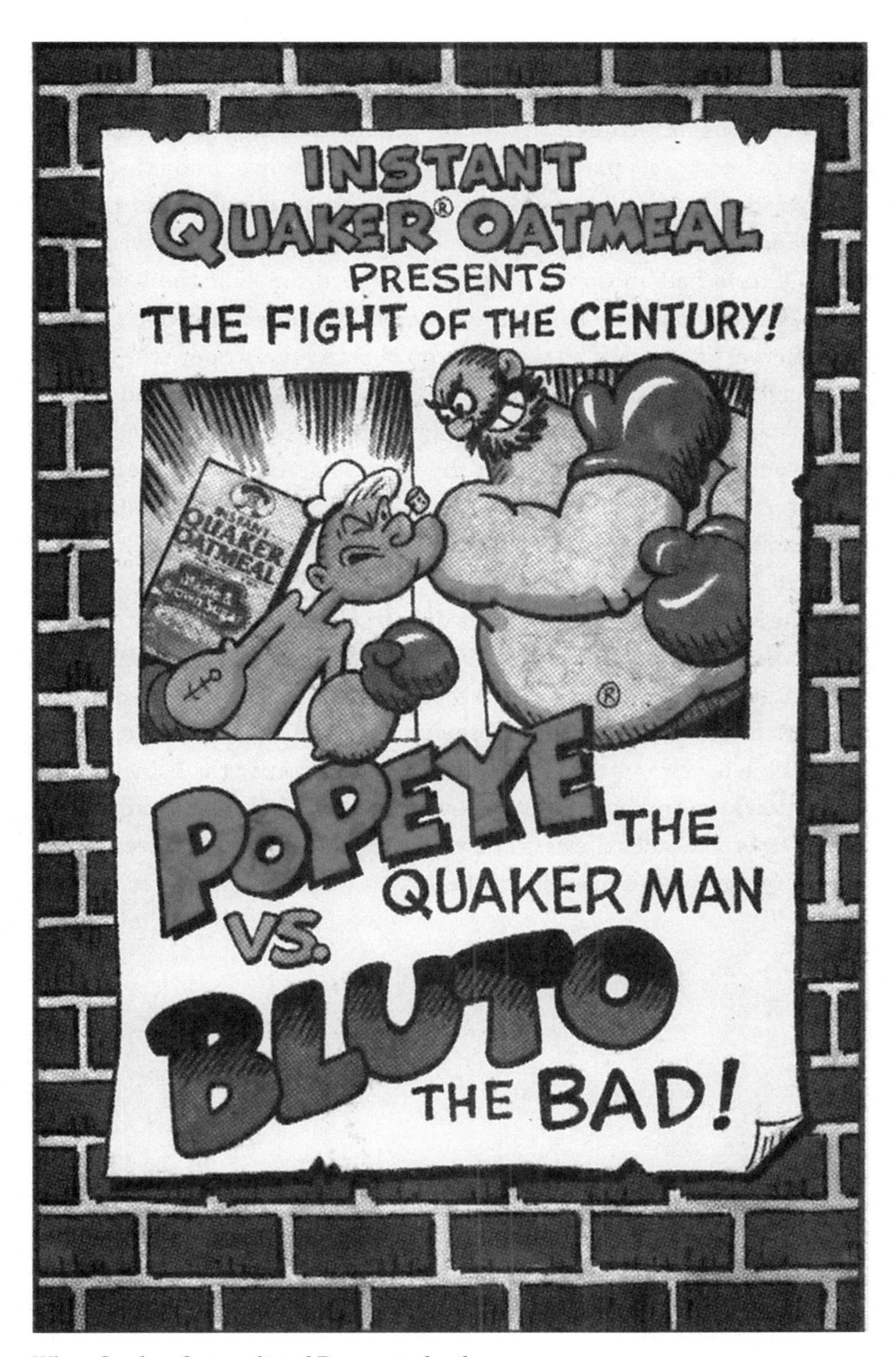

When Quaker Oats enlisted Popeye to be the new mascot for its instant oatmeal line in 1989, the company was unprepared for the hostile reaction of the Quaker church. It seems the famously nonviolent congregation objected violently to the nomenclature of "Popeye the Quaker Man."

tube by stating in Popeye's slogan, "Quaker Man" referred to the product he was eating and not his religious affiliation, and the former claiming, "To portray us as a church that beats people up is not acceptable." The general public, for its part, sided with neither, simply opining that it was a stupid idea to have Popeye reject his long-beloved spinach in favor of sugar-and-calorie-laden instant oatmeal. The Quaker ad executives might possibly have had an opportunity to hear recordings of the Wheatena shows, but obviously most of the company's customers had not.

In this particular case, the church won out, and the "Popeye the Quaker Man" campaign was quietly swept under the proverbial rug and forgotten. It was somewhat ironic that if the idea had been presented only a decade earlier, it would have been shot down because of the NAB's code against showing a product endowing its consumer with super strength, but in 1989, it was truly a David-and-Goliath scenario in every sense of the term.

In the big picture, though, it was only the latest in a series of strange adventures befalling the cereal advertising industry over the previous century. Regardless of what sort of regulations have been placed on advertisers or subsequently lifted, or what sort of mercenary attitude was responsible for either, those of us who grew up during the baby boomer generation know many of those beloved cereal mascots were some of our first friends. Whether they are still in use or have been long since retired, we still retain a warm place in our hearts (and our stomachs) for them.

Illustration Credits

Unless otherwise indicated, all images are from the author's personal collection.

Todd Franklin: pages 18, 20, 111, and color page 9 (*bottom*)

Dan Goodsell: pages 77, 80, 93, 94, 103, 118, 138, 147, 151, 157, 165, 168, and color pages 10 (*top*), 14 (*top and bottom*), 15 (bottom)

Donnie Pitchford: pages 81, 84, 101, 129, 132, 133, page 142, 152, 160, and color pages 5, 6 (*top*), 7 (*bottom*), 8 (*bottom*), 13 (*bottom*)

Bibliography

Argyropoulos, Paul. *The Wacky Packages Gallery*. Self-published, 2000.

Arnold, Mark. *Total TeleVision Productions: The Story of Underdog, Tennessee Tuxedo and the Rest*. Albany, Ga.: BearManor Media, 2009.

Bruce, Scott. *Cereal Box Bonanza: The 1950s*. Paducah, Ky.: Collector Books, 1995.

———. *Cereal Boxes and Prizes: The 1960s*. Concord, Mass.: Flake World, 1998.

———. *Cerealizing America: The Unsweetened Story of American Breakfast Cereal*. Winchester, Mass.: Faber and Faber, 1995.

Burke, Timothy, and Kevin Burke. *Saturday Morning Fever: Growing Up with Cartoon Culture*. New York: St. Martin's Griffin, 1999.

Dotz, Warren. *Advertising Character Collectibles*. Paducah, Ky.: Collector Books, 1993.

Dunning, John. *On the Air: The Encyclopedia of Old-Time Radio*. New York: Oxford University Press, 1998.

Efron, Edith. "The Children's Crusade that Failed." Three-part series. *TV Guide*, April 7, April 14, and April 21, 1973.

Grossman, Gary. *Saturday Morning TV*. New York: Dell, 1981.

Hall, Jim. *Mighty Minutes: An Illustrated History of Television's Best Commercials*. New York: Harmony Books, 1984.

Heighton, Elizabeth J., and Don R. Cunningham. *Advertising in the Broadcast Media*. Belmont, Wash.: Wadsworth, 1976.

Hollis, Tim. *Ain't That a Knee-Slapper: Rural Comedy in the Twentieth Century*. Jackson: University Press of Mississippi, 2008.

Hollis, Tim, and Greg Ehrbar. *Mouse Tracks: The Story of Walt Disney Records*. Jackson: University Press of Mississippi, 2006.

Lawson, Tim, and Alisa Persons. *The Magic behind the Voices: A Who's Who of Cartoon Voice Actors*. Jackson: University Press of Mississippi, 2004.

Lesser, Robert. *A Celebration of Comic Art and Memorabilia*. New York: Hawthorn Books, 1975.

Lester, Robie. *Lingerie for Hookers in the Snow: An Audiography of a Voice Artist*. Albany, Ga.: BearManor Media, 2006.

Margerum, Eileen. "The Case for Sunny Jim: An Advertising Legend Revisited." *Sextant: The Journal of Salem State College* (Fall 2001/Spring 2002).

Marquette, Arthur F. *Brands, Trademarks and Good Will: The Story of the Quaker Oats Company*. New York: McGraw-Hill, 1967.

Morgan, Hal. *Symbols of America*. New York: Viking Penguin, 1986.

Ohmart, Ben. *Welcome, Foolish Mortals: The Life and Voices of Paul Frees*. Boalsburg, Penn.: BearManor Media, 2004.
Ohmart, Ben, and Joe Bevilacqua. *Daws Butler: Characters Actor*. Boalsburg, Penn.: BearManor Media, 2005.
Roden, Steve, and Dan Goodsell. *Krazy Kids' Food: Vintage Food Graphics*. Los Angeles: Taschen, 2003.
Sagendorf, Bud. *Popeye: The First Fifty Years*. New York: Workman, 1979.
Scott, Keith. *The Moose That Roared*. New York: St. Martin's Press, 2000.
Woolery, George. *Children's Television: The First Thirty-Five Years, 1946–1981*. Metuchen, N.J.: Scarecrow Press, 1985.

Pinning down sources where interested parties may view the actual commercials described in this book can be a bit tricky. As mentioned in the preface, dozens of unlicensed video compilations have been released, dating back to the days of VHS tape, and most of them would be impossible to find today. However, there are many channels on the YouTube website where vintage cereal commercials can be accessed. At least as of the time of this writing, a visit to any of the following YouTube channels will provide a rich overview of these classic TV spots:

CartoonBrew
VintageTVCommercials
Pretzel78
CountChoculatte
Muttley16
TherealRNO
Freenbean
Toontracker
Genius7277
Allcommercials

Searching YouTube for the name of any character or cereal discussed in this book will produce a multitude of results—surely the quickest and most reliable way to view examples of a particular ad campaign's evolution.

In addition, Duke University is now the custodian of the archives of the Benton and Bowles advertising agency, which handled the General Foods/Post account for decades. Many of the commercials produced for the company's products, from Sugar Crisp to Pebbles, can be downloaded and watched at the website http://library.duke.edu/digitalcollections/.

Index

Page numbers in italics refer to illustrations

TIM HOLLIS is the author of twenty-one books chronicling various aspects of popular culture and history, including *Hi There, Boys and Girls! America's Local Children's TV Programs*, *Mouse Tracks: The Story of Walt Disney Records*, *Selling the Sunshine State: A Celebration of Florida Tourism Advertising*, and *Ain't That a Knee-Slapper: Rural Comedy in the Twentieth Century*.

The University Press of Florida is the scholarly publishing agency for the State University System of Florida, comprising Florida A&M University, Florida Atlantic University, Florida Gulf Coast University, Florida International University, Florida State University, New College of Florida, University of Central Florida, University of Florida, University of North Florida, University of South Florida, and University of West Florida.